TABLE OF CONTENTS

GLOBAL WARMING

THE BENEFITS
OF EMISSION ABATEMENT

ORGANISATION FOR ECONOMIC CO-OPERATION AND DEVELOPMENT

ORGANISATION FOR ECONOMIC CO-OPERATION AND DEVELOPMENT

Pursuant to Article 1 of the Convention signed in Paris on 14th December 1960, and which came into force on 30th September 1961, the Organisation for Economic Co-operation and Development (OECD) shall promote policies designed:

- to achieve the highest sustainable economic growth and employment and a rising standard of living in Member countries, while maintaining financial stability, and thus to contribute to the development of the world economy;
- to contribute to sound economic expansion in Member as well as non-member countries in the process of economic development; and
- to contribute to the expansion of world trade on a multilateral, non-discriminatory basis in accordance with international obligations.

The original Member countries of the OECD are Austria, Belgium, Canada, Denmark, France, Germany, Greece, Iceland, Ireland, Italy, Luxembourg, the Netherlands, Norway, Portugal, Spain, Sweden, Switzerland, Turkey, the United Kingdom and the United States. The following countries became Members subsequently through accession at the dates indicated hereafter: Japan (28th April 1964), Finland (28th January 1969), Australia (7th June 1971) and New Zealand (29th May 1973). The Commission of the European Communities takes part in the work of the OECD (Article 13 of the OECD Convention). Yugoslavia has a special status at OECD (agreement of 28th October 1961).

Publié en français sous le titre :

RÉCHAUFFEMENT PLANÉTAIRE
Les avantages de la réduction des émissions

FOREWORD

This report was prepared under the 1991 "Socio-Economic Aspects of Climate Change" activity of the OECD Environment Committee. The Committee recommended that it be made available to the public on the responsibility of the Secretary General, who subsequently agreed. It does not necessarily reflect the views of the OECD or its Member countries.

ALSO AVAILABLE

Climate Change: Evaluating the Socio-Economic Impacts (1991)
(97 90 02 1) ISBN 92-64-13462-X   FF130   £16.00   US$28.00   DM50
Environmental Policy Benefits: Monetary Valuation (1989)
(97 88 07 1) ISBN 92-64-13182-5   FF95   £11.50   US$20.00   DM39
Responding to Climate Change. Selected Economic Issues (1991)
(97 91 04 1) ISBN 92-64-13565-0   FF150   £21.00   US$36.00 DM62

Prices charged at the OECD Bookshop.
THE OECD CATALOGUE OF PUBLICATIONS and supplements will be sent free of charge
on request addressed either to OECD Publications Service,
or to the OECD Distributor in your country.

PREFACE

At whatever level it may be considered, global warming is arguably the most challenging problem facing environmental policy-makers today. Certainly, it is among the most complex. Not only does it involve all countries in the world, it also encompasses a wide range of emission types, sources, and sinks. To further complicate matters, each of these variables must be viewed over long periods of time, involving several generations.

All analysts attempt to simplify this complex array of variables using some structured model. The economist is no different from the chemist or the biologist in this regard. To an economist, the most appropriate analytical structure will usually be some variant of the "benefit-cost" model. This is because a major concern of economists is to develop policies that are economically-*efficient* in social terms. This does not mean that economic efficiency is the only, or even necessarily the most important, policy objective in the global warming debate. But it will be one important element in that discussion, and understanding economic efficiency implies being able to say something about both the costs and the benefits of proposed response strategies.

Once it is accepted that policy decisions should be based (at least partially) on benefit and cost considerations, the next task is to decide how big those benefits and costs are likely to be, and in what economic sectors they will likely occur. In terms of estimating the economic *costs* of responding to climate change, many models already exist to simplify the task of arriving at credible estimates. Unfortunately, in the case of estimating the economic *benefits* of response policies (the "proxy" for which is the environmental damage avoided by those policies), very few guidelines presently exist to help economists develop credible estimates.

And yet, it may be precisely the interpretation of these benefits that will guide society's collective response to the threat of global warming. If the benefits appear to be small, the level of costs that can be justified to react to the problem will also be small, and society's willingness to abate emissions, or to otherwise adapt to the problem, will be correspondingly limited. In such a case, inaction will usually appear to be the most efficient policy response.

In some ways, this conclusion is already being reached in some parts of the global warming debate. Some countries feel they will actually *benefit* from global warming. Others perceive that global warming will only impact on certain (low cost) economic sectors. Still others feel that, even where the impacts are likely to be significant, the costs of preventing those impacts are likely to be even higher. In all three situations, inaction becomes the logical policy response (at least from an economic efficiency perspective).

But what if the benefits of responding to global warming are being undervalued by these analysts? Valuing the environment is probably the most difficult theoretical prob-

7

lem facing an economist, and considerable debate surrounds any attempts to do so. Nevertheless, OECD has been actively involved in this problem for many years. Given the importance of the climate change issue, it seemed appropriate for OECD to attempt to apply its experience to that issue.

In this report, Dr. William Cline makes two important contributions to estimating the benefits of global warming response policies.

First, he provides a framework for structuring the analysis. The structure outlined in Section 2 is designed to be *comprehensive,* in the sense that not only the direct economic consequences of global warming are enumerated. Also included are indirect economic impacts (e.g. price effects) and environmental impacts. Equally importantly, Dr. Cline's model is strongly based on the underlying science of the greenhouse effect, and on the *very long term* impacts that present and future emissions will create. This emphasis on the very long term is especially important, given that most analyses of economic impacts that have been done to date have focussed on the ''2XCO$_2$'' scenario that formed the basis of work done by the Intergovernmental Panel on Climate Change (IPCC).

In the second part of the report, Dr. Cline attempts to provide rough estimates of values for the most significant cells in this conceptual framework. Although these estimates are clearly very crude and preliminary, the emphasis that the author places on the very long term in arriving at these numbers gives considerable pause for thought.

If he is right, the value of the benefits of responding to climate change may be of the same ''order of magnitude'' as most presently-available estimates of the economic costs of policy action.

The tantalizing conclusion emerges that more abatement of greenhouse gas emissions may be justified than is presently accepted in much of the available literature. To better understand the robustness of this conclusion, more research of these variables is clearly in order, and the report concludes with several suggestions in that direction.

GLOBAL WARMING:
ESTIMATING THE ECONOMIC BENEFITS OF ABATEMENT[1]

by

William R. Cline*

* Dr. Cline is a Senior Fellow at the Institute for International Economics, Washington, D.C., USA. His research interests encompass a wide variety of international environmental topics, particularly those involving agricultural, trade, and developing country themes.

INTRODUCTION

It has become increasingly clear that international action on the greenhouse effect is unlikely to be undertaken solely on the basis of scientific opinion that global warming will take place. The Bush Administration, in particular, has cast the issue as one of balancing environmental concerns on the one hand, and economic costs on the other. At the White House Conference on Science and Economics Research Related to Global Change in April, 1990, US spokesmen emphasized the scientific uncertainties associated with the greenhouse effect and the burdensome economic costs of curtailing carbon emissions, especially for developing countries. At the first meeting of the UN International Negotiating Committee for a Framework Convention on Climate Change (Chantilly, Virginia, February 1991), US officials for the first time acknowledged the greenhouse was a "potential threat" that justified "taking action now". However, the new US position offered only to stabilize trace gas emissions through the year 2000 (not thereafter), and to do so almost entirely through phasing out CFCs – a commitment already made under the Montreal Protocol undertaken internationally, in response to the danger of depletion of the stratospheric ozone layer. Unlike many other countries, the US was unprepared at this time to offer limitations on carbon dioxide emissions, and indeed the expectation was that these would rise by a further 15 per cent by the year 2000 (Kerr, 1991).

The US Administration could not be faulted from the standpoint of public policy in seeking to balance the costs and benefits of action to reduce global warming. Indeed, its position – to phase out CFCs but to go no further – was close to what one well-known analyst has advocated as the extent of action that is justified on a cost-benefit basis (Nordhaus, 1990). The problem is that the empirical basis for this cost-benefit judgment is extremely slim. Several studies have been developed on the cost side, typically through the use of energy-economic models (Manne and Richels, 1990a; Edmonds and Barnes, 1991; Jorgensen and Wilcoxen, 1990; Congressional Budget Office, 1990; Nordhaus, 1990; for surveys, see Hoeller, Dean, and Nicholaisen, 1990; Edmonds and Barnes, 1991; Darmstadter, 1991). On the benefits side, however, the estimates are scant, tend to be formulated only for a few individual effects rather than calculated on a comprehensive basis, and tend to give qualitative descriptions of damage from global warming, rather than quantitative estimates of the corresponding economic magnitudes (IPCC, 1990b). Until more adequate economic analysis of abatement benefits becomes available, there may be a considerable tendency in the policy-making process toward bias against action simply because the costs have been much more fully spelled out than the prospective gains.

This essay seeks first to set forth a conceptual framework for analyzing the benefits side of policy action on the greenhouse effect. The focus will be on abatement of warming through trace gas emissions limitations, although adaptation is also explicitly considered as part of the model. Within this conceptual framework, the discussion emphasizes that global warming will affect each country differently, and in varying intensity. Climate change will also have impacts on society that are difficult to capture using traditional economic measures such as GNP. "Environmental variables", such as species diversity, human amenity, and morbidity are especially important in this regard. Other important dimensions of the impacts include their timing, the level of policy response (both abatement and adaptation), risk aversion, and equity. The methodological approach described in the first part of the paper considers each of these aspects.

After describing the conceptual framework, the discussion turns to a survey of existing literature as a point of departure for rough estimates of prospective damages from global warming. The analysis then develops new estimates of potential prospective damages from global warming (primarily for the case of the United States) for each major category of impact. It is important to keep in mind that these "ballpark" estimates are very preliminary. Nonetheless, an important implication of the calculations is that the potential impacts of climate change are probably more significant than those suggested by most previous studies. This is especially true if the very-long term is considered, and if damages outside the normal economic system are taken into account.

The paper closes with an outline of future research that should be undertaken to advance our understanding of the potential benefits of global warming response policies. In view of the strong emphasis that has been placed to date on researching the *costs* of these policies, it is now time to place greater emphasis on the examination of their *benefits*. This shift in focus should ultimately provide more reliable benefit-cost analysis of greenhouse policy.

It should be emphasized at the outset that the paper's perspective on global warming stresses that policy must take account of effects over an extremely long time horizon. Most of the analysis has been specified in terms of the equilibrium effects of a doubling of atmospheric carbon-dioxide trace gas equivalent. However, that benchmark is expected to have already been reached by the year 2025 (IPCC, 1990*a*), although ocean thermal lag is expected to delay the full consequential warming by some three decades or so. Unfortunately, the problem does not stop with a doubling of carbon-dioxide. In the absence of specific policies, the build-up may be expected to continue for perhaps three centuries. It is estimated here that, after taking account of economic and emission trends and giving explicit consideration to the economically-available fossil fuel resource base, by the year 2250, the globe could be committed to greenhouse warming that is several times as large as that associated with CO_2-equivalent-doubling. Specifically, if we define λ as the "climate sensitivity" parameter [i.e. the mean global warming to be expected from the doubling of CO_2 (equivalent, all trace gases)], the present central estimate is that $\lambda = 2.5°C$ (IPCC, 1990*a*). By the year 2250, the commitment to global warming based on the same parameter would be a central estimate of 10°C (Cline, 1990*b*).

Incorporation of long time horizons is important if the cost-benefit calculus is to be balanced. The costs of suppressing fossil fuel use and carbon emissions, as well as other trace gas emissions, will be evident early in the time path of any policy action program, whereas the warming thereby avoided pertains to periods decades later, and the damage may be expected to rise more than linearly with temperature. Inclusion of distant horizons in turn makes the issue of time discounting central to the analysis.

CONCEPTUAL FRAMEWORK

The benefits obtained from a policy of greenhouse warming abatement are the damages avoided as a consequence of the action, net of any economic gains that might have been encountered under larger global warming, and taking due account of adaptation alternatives. Calculation of the benefits thus requires estimation of a future baseline of net damage from global warming, as well as calculation of the alternative time path of net damage under the abatement policy scenario under consideration. The benefits of the policy then equal the difference between the baseline and the scenario damage time paths. The value of this stream of damage reduction may then be compared with the costs of the policy, arising from the shifts to higher-cost alternatives: lower-carbon fossil fuels (e.g. natural gas); sustainable biomass in place of traditional fossil fuels; non-fossil-fuel energy (e.g. nuclear, solar); alternative product mixes with lower energy content; and greater use of other factors of production (capital, labor) as replacements for energy in the production process.

Categories of economic effects

The first task is to identify the categories of damage from global warming, and thus the areas of potential benefits from policy action. Table 1 lists seventeen areas of economic effects, and may well be incomplete. The impacts on agriculture and sea level rise have received the most attention. The table separates agricultural production from agricultural consumption, because a country can be seriously affected by global reductions in agriculture even (indeed, especially) if it produces no food domestically. Forest loss is anticipated as the rate of migration of forests is expected to be unable to keep pace with the rate of temperature change. For the same reason, it is expected that some species will be lost. The impacts in the area of ecosystems raise special problems for economic valuation.

The next three economic effects concern the damage from sea level rise. In coastal urban areas, it is likely that seawalls and dikes would be built. Developed islands have sufficiently valuable real estate that their surface level might be raised, using imported sand. The buildup of seawalls and other barriers would tend to cut off estuaries and thus eliminate wetlands, an important resource for wildlife. Where barriers are not built, some coastal dryland would be sacrificed.

The next two economic effects concern the impact of a warmer climate on man's existing climate-interference services: air conditioning and heating of homes, office buildings, and other structures. A warmer climate would require greater energy inputs

13

Table 1. Desireable Dimensions for Analysing the Benefits of Greenhouse Warming Emission Abatement

Primary Dimensions

A. Economic Effect

1. Agricultural Production	10. Human Amenity
2. Agricultural Consumption	11. Human Life and Morbidity
3. Forest Loss	12. Storm and Hurricane Loss
4. Species Loss	13. Construction
5. Coastal Defences	14. Leisure Activities and Tourism
6. Wetlands Loss	15. Water Supply; Non-agricultural
7. Coastal Drylands Loss	Drought
8. Space Cooling	16. Urban Sanitation
9. Space Heating	17. Pollution Abatement Spillover

B. Geographical Region

1. US-East	16. Brazil
2. US-South	17. Indonesia
3. US – Central North	18. Bangladesh
4. US – Central Plains	19. Other Southeast Asia
5. US – Southwest	20. Other East Asia
6. US – Northwest	21. Mexico, Central America
7. Canada	22. Other South America, Low Latitude
8. EC – Mediterranean	23. Other Southe America, High Latitude
9. EC – France, Germany, UK, Benelux	24. Middle East
10. Scandinavia	25. Sahel
11. Eastern Europe	26. Other Africa, Mid Latitude
12. Former USSR Europe – North	27. Other Africa, High Latitude
13. Former USSR Europe – South	28. Australia, New Zealand
14. Former USSR – Asia	29. Island Nations
15. China	

C. Time/Warming Horizon

1. CO_2-equivalent doubling, equilibrium ($2^{1}/_{2}°C$)
2. Very-long-term (250 years, 10°C)

Secondary Dimensions

A. Baseline Warming

1. Low
2. Medium
3. High

B. Policy Action Scenario

1. Moderate Action
2. Severe Action

C. Discount Rate

1. Zero
2. Low-intermediate
3. High-market

D. Risk Aversion

1. Neutral
2. Risk Averse

E. Weighting

1. Market Weights
2. Equity Weights

(and equipment) for space cooling, but with some savings in activity currently devoted to space heating.

A warmer climate would affect both the amenity and the salubrity of human life. In cold climates, these may improve. Elsewhere, there may be deterioration. The disamenity of increased frequency of heat waves, and possible increased incidence of heat-related deaths and disease, are among the damages to be expected from warming. The severity of these effects becomes far greater with the higher magnitudes of very-long-term warming.

Migration costs are closely related to the category of human amenity. As discussed below, migration forced by warming would represent a cost to the migrant who otherwise would prefer to stay in his native area. At the same time, if warming caused dislocation in developing countries and increased migration (legal and illegal) toward the industrial countries, there could be costs to the host countries in the form of increased social infrastructure requirements, and even costs of increased political instability and social conflict.

Global warming is expected by many to increase the frequency and severity of tropical cyclones (as discussed below), and perhaps other storm damage. Warming could also affect the construction industry, in both positive and negative ways. Global warming is expected to increase precipitation, and construction is adversely affected by rain. However, where cold weather is a constraint on construction, warming could have a favorable effect.

Warming would tend to have an adverse effect on the ski industry. If ocean warming bleaches coral reefs, there could be adverse effects on the tourist industry. Some offsets might occur in other leisure activities, such as increased winter camping.

As discussed below, there are indications that warming could cause a severe increase in the incidence of droughts. In principle, the agricultural impact of this effect would already be incorporated in the estimates for the agricultural sector. However, there are non-agricultural effects of drought as well, and in particular increased costs of water supply for household and industrial use.

Urban sanitation could experience adverse effects in some coastal cities where sea level rise would damage existing infrastructure.

The final economic effect listed in the table is qualitatively different from the rest. It concerns the spillover effects in other areas of pollution that arise from greenhouse policy. Thus, if policies to reduce carbon emissions are undertaken for anti-greenhouse purposes, the result is likely to be other reductions in air pollution (e.g. lower automobile emissions resulting from reduced gasoline burning). This same category could potentially include major elements working in the opposite direction. The release of dust into the atmosphere, or the iron fertilization of the oceans as "geo-engineering" measures to counter the greenhouse effect, could have severe negative environmental side effects.

Inherent limits to damage?

Before turning to the additional dimensions of the analysis of abatement benefits, it is useful to consider an argument that stems solely from the "economic categories" dimension. Nordhaus (1990) and Schelling (1990) have argued that, at least for industrial countries, the potential damage from global warming is greatly limited by the small share of climate-sensitive sectors in the structure of Gross Domestic Product (GDP). For example, with agriculture contributing less than 2 per cent of GDP, even sharp reductions in agricultural output would represent relatively small percentages of GDP. There is obviously some merit in this point, as one would expect a developing economy with a far higher share of agriculture to be more vulnerable than an industrial economy. However, the argument may tend to understate potential damages for industrial economies.

Of the seventeen categories identified in Table 1, only six are directly linked to the size of specific productive sectors in the economy (agricultural production, space cooling and heating, construction, leisure activities, and sanitation). There is no *a priori* linkage of prospective damage in the other categories to a particular sectoral economic base. Moreover, of these other categories, some are more likely to show a higher impact in an industrial economy than in a developing economy. Human life is one such category, if standard approaches based on lifetime earnings (for example) are the criterion. Coastal defenses are likely to be another: the prospective investment in defending New York or Amsterdam is likely to exceed that of defending Recife, Brazil (abstracting from different physical situations), because of higher structural build-up and property valuation in the more developed country cases. The same may be said of the "pollution spillover" category.

Even in agriculture, the "inherent limitation" argument ignores the diamond-water paradox. Agriculture is only three per cent of US GDP, in part because agricultural production is abundant and prices are low. With serious cutbacks in agricultural production, and given price-inelastic demand, *ex poste* prices would be much higher. Proper evaluation of the loss of consumer surplus (hence Category "2") could mean a much larger damage potential than might be suspected from the small share of agriculture in present GDP. In sum, it is misleading to cite the low shares of climate-sensitive sectors in the existing structure of production as *prima facie* evidence that the benefits of global warming abatement are likely to be small.

Geographical detail

Table 1 presents a list of world areas that are appropriate for evaluation of the effects of global warming. The list is based on important differentiation by latitude, a key element in expected differential warming; geopolitical weight; and major actors in the policy response. Thus, China, Brazil, and Indonesia warrant separate treatment because of their unique roles in coal resources or deforestation. Bangladesh merits separate consideration because of its high vulnerability to sea level rise.

It is common to assert that regional analysis is beyond the capability of the General Circulation Models (GCMs) used for global warming projections. Actually, the more recent high-resolution models provide sufficiently detailed grids that they are encompassed within geographical areas of the size shown in Table 1. Thus, the United Kingdom Meterological Office model, with a grid resolution of 2.5° horizontal and 3.75° vertical, divides the earth's surface into 6 912 grids (360/2.5=144 horizontal 180/3.75=48 vertical). The vertical distance of the grid is 417 km, about the distance from Washington D.C. to New York. At the equator, the horizontal distance is 278 km; at 40° latitude (N or S), it is 213 km (Cline, 1990c). IPCC Figure 1 reports the average of three high-resolution models' estimates for mean winter warming by this grid resolution (Cline,

16

1990*b,* based on IPCC, 1990*a*). It is evident that the detail is sufficient for discrimination among the geographical areas suggested in Table 1[2].

Whether the high-resolution models are reliable is, of course, another question. The Intergovernmental Panel on Climate Change (IPCC) finds that the three models referred to in Figure 1 (UK Meterological Office, UKMO; Canadian Climate Center, CCC; and Geophysical Fluid Dynamics Laboratory, GFDL) are relatively accurate in simulting present-day climate. Nonetheless, it warns that for regional calculations, "confidence in these estimates is low, especially for the changes in precipitation and soil moisture." (IPCC, 1990*a*, p. 163, 165). Because grid size is sufficiently small, it is evident that the limited confidence is not the result of inadequately detailed grid resolution, so the implication is that there is limited confidence in the projections, even for whole blocks of grids constituting a wide area[3].

For policy analysis, there is little alternative to using the detailed GCM estimates as the point of departure, with the understanding that they are central estimates with potentially wide bands of uncertainty. The emerging body of agricultural estimates (discussed below) does just that, as the estimates typically accept the GCM projections for a particular area, and then apply crop models to calculate the impact.

There is some risk that the policy process could be impaired by regional estimates, as some countries might turn out to be "winners", and thus be unprepared to participate in international abatement efforts. This risk is probably more than outweighed by the fact that, before nations are likely to undertake costly abatement measures, their governments are going to insist on a better understanding of the stakes in the economic benefits, and some regional detail will be essential to this assessment.

Damage and benefits matrices

The economic effects and geographical areas listed in Table 1 provide a framework for assessing the benefits of policy action. Let **D** be a matrix of expected economic damage, with "m" rows for the various economic effect categories and "n" columns for the individual geographical regions to be examined. Let subscript "b" refer to the baseline – the situation in the absence of any special policy (although the Montreal Protocol on CFCs should be included as part of the baseline). Let subscript "p" refer to the policy scenario being considered. Let subscript "t" refer to the time period. Then the benefits at a particular benchmark year in the future, t, from the policy program will equal:

[1] $$\mathbf{B}_{p,t} = \mathbf{D}_{b,t} - \mathbf{D}_{p,t}$$
$$\text{mxn} \quad \text{mxn} \quad \text{mxn}$$

where **B** is the matrix of benefits by economic effect and geographical region. Global benefits ($B_{G,t}$) for the year in question are thus:

[2] $$\mathbf{B}_{G,t} = \sum_i \sum_j b(i,j)_{p,t}$$

where $b(i,j)_{p,t}$ is the cell in row "i" and column "j" of the benefits matrix[4].

An important feature of this approach is that it recognizes at the outset that it may be impossible to avoid all greenhouse damage. The total extent of the benefits of greenhouse warming abatement is thus likely to be less than the outright measure of the damage of

17

global warming. That is, for the benefits of action to equal simply the total projected damage, the policy action would have to be capable of completely eliminating global warming and its consequences. Only in that case would the policy scenario damage matrix, $D_{p,t}$, be a set of zeros, such that the benefits matrix B would equal the full value of the baseline damage matrix, $D_{b,t}$.

This point has been emphasized by some economists, in what may be called the "mere delay" school of opposition to greenhouse policy action. This school received some early support from simulations seeming to show that even aggressive policies would do little to affect the greenhouse effect. Edmonds and Reilly thus calculated that taxes ranging from 56 per cent on gas up to 100 per cent on coal and 115 per cent on shale oil (differentiated by fuel carbon emissions) would only delay the date of doubling of atmospheric carbon dioxide by 10 years (Edmonds and Reilly, 1983). However, subsequent simulations with the same model showed much greater impact, largely because of a less pessimistic baseline (Cline, 1989).

Even so, the "mere delay" school tends to argue that, whereas policy might postpone global warming, it cannot avoid the eventual warming. The adherents of this school thus emphasize that economic evaluation must consider only the benefits of delaying greenhouse damage, rather than calculating that it can be avoided altogether. Cline (1990*d*) examined this issue for the very long term, and estimated that whereas the commitment to global warming by the year 2250 is likely to be 10.6°C in the base case of no policy intervention, it could be held to 5.5°C by a freeze in emissions at present absolute levels, and to as low as 2.5°C under the IPCC's "accelerated policies" scenario, the most aggressive considered by that group (IPCC, 1990*a,* p. 349)[5]. These calculations of warming even with policy action are sobering, since they suggest that perhaps the best that can be hoped for is to limit long-term damage to what has so far been associated with the conventional benchmark of the doubling of carbon dioxide. The central point is that, in the absence of action, the warming and damage could be much worse, so that the difference between $D_{b,t}$ and $D_{p,t}$ may be large, even if the latter is unavoidably large.

As Table 1 indicates, there are other dimensions to the analysis beyond "economic effect" and "geographical area". The third "primary" dimension is the time horizon. As suggested in the Table, it is important to specify at least two benchmark periods. A first, obvious, period is that associated with the equilibrium effects of a doubling of carbon dioxide-equivalent trace gases, in view of the large portion of greenhouse analysis that is premised on this benchmark. As noted above, in the "business as usual" scenario, this doubling could occur as early as 2025. With thermal ocean lags, the equilibrium effects would be felt by perhaps 2050. As discussed above, a benchmark period should include the "very-long-term." For reasons presented later, the commitment year 2250 (and the related equilibrium-impact year of perhaps 2280) is proposed as being appropriate for this horizon. This period is of particular relevance because it is still within a time frame when fossil fuel resources remain available at economically feasible extraction costs, and it is the beginning of a threshold period when deep ocean mixing of carbon dioxide begins to offer the possibility of significant reversal of atmospheric buildup (Cline, 1990*d*; Sundquist, 1990).

Ideally, the time dimension would include intervening periods as well, perhaps at 25-year intervals. In the absence of the greater temporal detail, the path of annual benefits may be approximated by some form of interpolation between the two benchmark years. Straight-line interpolation is likely to overstate the early benefits, because the damage from global warming is likely to be non-linear (rising more than proportionately with the cumulative temperature change).

One possible assumption is merely that the damage experiences steady compound growth between the first benchmark (CO_2-doubling) and the second (very-long-term, e.g. 2250 commitment, 2280 equilibrium). Designating "d" as the former and "L" as the latter years, for a given cell in the benefits matrix, the estimated benefit at an interim year "t" may be estimated as:

[3] $b(i,j)_{p,t} = b(i,j)_{p,d}\, e^{r(i,j)t}$; $d < t < L$

where

[4] $r(i,j) = \{\ln[b(i,j)_{p,L}] - \ln[b(i,j)_{p,d}]\}/\{L-d\}$

and "e" refers to the base of the natural logarithm (ln).

For years prior to the first benchmark, the same approach suggests "backward growth" at the rate identified for the inter-benchmark period. However, as the growth rate could be low, this result could leave the estimated benefit in early years relatively high (close to that of the CO_2-doubling benchmark), even though, by definition, at the beginning of the period (e.g. starting in the early 1990s), the warming damage (and abatement benefit) is zero. To assure convergence to zero initial benefit, the benefit in a year prior to the CO_2-doubling benchmark may be estimated as:

[5] $b(i,j)_{p,t} = $ Min: $\begin{vmatrix} [t/d]\ b(i,j)_{p,d} \\ b(i,j)_{p,d}/e^{r(i,j)[d-t]} \end{vmatrix}$; $t < d$

where the first alternative is simply linear interpolation, and the second is compound "backward growth".

Close analysis of the time profile of the benefits is important. The simplifying assumption of compound growth from the first benchmark to the second will tend to give relatively low estimates in the early part of the period, less than under linear interpolation. Considering that the underlying population growth and emissions trajectories are premised on stabilization of population in the mid-21st century, the actual path might show an earlier increase, perhaps similar to a logistics curve, rather than one of geometric growth.

If it can be established that the "damage function" follows a particular geometric relationship with respect to the magnitude of global warming, then an improved estimate of the time path of benefits that replaces equations [3] through [5] may be calculated as:

[5'] $b(i,j)_{p,t} = b(i,j)_{p,d}[\Delta T_t/\lambda]^\gamma$

where ΔT_t is the amount of global warming by year t (degrees celsius), λ is the amount of warming associated with benchmark doubling of carbon dioxide equivalent (the climate sensitivity parameter), and γ is the exponent in the geometric relationship of damage to warming.

For purposes of estimating ΔT_t, linear interpolation between benchmark $2xCO_2$ and "very-long term" warming should be approximately valid. The reason is that with the expansion path of economic activity, carbon emissions, and atmospheric concentration of

19

trace gases more than linear with respect to time, combined with a less than linear (i.e. logarithmic) relationship of radiative forcing, and thus warming, with respect to carbon dioxide concentrations, one should expect the resulting warming to be approximately linear with respect to time (Cline, 1990*b*). Equation [5'] should apply both before and after the year associated with benchmark carbon-equivalent doubling.

Scaling

Estimation of the baseline damage and the policy scenario damage requires some means of scaling the initial absolute estimates. The typical absolute estimates are based on the impact that the warming would have on the present-day economy. Thus, agricultural estimates tend to be based on yield impacts as applied to the current production base. However, the scale of the base may be expected to grow dramatically over the long time horizons under consideration.

One approach to the scaling problem is to leave the primary analysis in absolute terms scaled to today's economic magnitudes, and then to distinguish between those effects that are likely to grow with the scale of the economy, those likely to remain unchanged in absolute size, and those likely to fall somewhere in between. Thus, agricultural effects would tend to grow with the size of the agricultural base, which may be expected to grow less rapidly than the overall economy because of Engel's Law (income-inelastic demand for food). In contrast, space conditioning would tend to grow proportionately (or even more than proportionately) with the economy (unitary or greater income elasticity of demand). Any effects that are "public goods", such as the valuation of a species, would plausibly grow proportionately with GDP, under the assumption that the public would be prepared to spend the same percentage of GDP on their preservation even at higher absolute GDP (and the growth might be more than proportionate to the extent that enjoyment of the environment is viewed as a luxury good, with income elasticity greater than unity).

In sum, scaling may be accomplished by first obtaining an absolute damage estimate against current economic scale, and then projecting to the future benchmark year on the basis of projected GDP growth, taking account of whether the effect is likely to maintain the same proportion to GDP, grow more slowly than GDP, or grow more rapidly. For GDP itself, the key elements for scaling are the assumptions on population growth and plausible rates of per capita income growth.

Adaptation

Schelling (1983, p. 453) has emphasized that analyses of greenhouse damage should not make the mistake of "superimposing a climate change that would occur gradually in the distant future on life as we know it today". There are two components to this admonition. The first concerns general changes in the economy. Although it is true that, for example, calculations in the early 19th century of future unavailability of whale oil would have overstated the problem because of the failure to anticipate the shift from candles to electric lights, in the absence of firm reasons to project specific technological breakthroughs, this type of "error" simply cannot be avoided. Nor is there any reason to think that the bias of these errors works in one direction or the other. Otherwise, the same argument could be used to contend that there was a systematic direction of bias in future climate analysis if the general expectation were that there would be a systematic cooling rather than warming of the earth, which would be equally untenable. In neither case can it be argued that "superimposing future change on today's life" introduces either an upward or a downward bias on damage estimates.

There is a second component of the argument, however, which can be directly incorporated into the analysis: those changes attributable to adaptation induced by global warming itself. For this set of changes, the proper method is to reduce the estimate of the damage by the change in the relevant comparison basis, on the one hand, while taking account of the costs of adaptation, on the other. If only the first is taken into account without the second, the result can be highly misleading. Consider an extreme warming that makes a particular location completely uninhabitable. If the "adapted" comparison basis is used, then there is no damage assessed, because there is no one in the location to be affected by the extreme warming. What is needed is to incorporate the costs incurred by the population forced to move out. As suggested by the fact that populations fight wars to remain in their traditional lands, these opportunity costs might be substantial.

This second effect of the changing economy – adaptation – may be taken into account in the benefits analysis as follows. Let $\Delta D_{a,t}$ (<0) be the change in the damage matrix associated with adaptive change, with "t" designating the year (as before). Let $C_{a,t}$ be the corresponding matrix of costs of carrying out the adaptation[6]. Then the estimate of abatement benefits may be adjusted for adaptation by:

[6] $\qquad B_{p',t} = D_{b,t} - D_{p,t} + C_{a,t} - (- \Delta D_{a,t})$

That is, the initial estimate of the benefit (equation [1]) can now add the benefit of avoiding the costs of adaptation, but must subtract the reduction in damage that could have been secured by adaptation. (The final double negaive arises because the change in damage resulting from adaptation is negative).

Discounting

As indicated in Table 1, in addition to the three primary dimensions of the benefits analysis (economic category, geographical area, and time horizon), there are at least five "secondary" dimensions. The discount rate for comparing values over time is one of the most important.

It is argued elsewhere that the appropriate rate is probably in the range of 1 to 2 per cent annually, rather than the more conventional 5 or even 10 per cent for project analysis (Cline, 1990b). The argument is based on three components. First, there should be no discount rate for "pure time preference in consumption": that is, no discounting solely for the earlier arrival of consumption.

Second, the discounting applicable for reasons of rising per capita consumption (and, hence, diminishing marginal utility is likely to be a very low rate); otherwise, the trade-off between extra consumption in 200 years and today becomes implausibly steep. More formally, it may be shown that under a logarithmic utility function (i.e. one in which utility rises as the logarithm of consumption), the appropriate rate of discount for consumption is simply the growth rate of per capita consumption (Cline, 1991c). Historically, this rate has been in the range of 1 per cent or less.

Third, although in principle, resources could be reallocated from consumption to investment and earn a higher rate in that use, society has demonstrated its unwillingness to do so quite apart from greenhouse matters, so the consumption/investment allocation of income should be taken as given. With that premise, the proper discount rate is a weighted average between a very low consumption discount rate, and a higher investment rate of return.

It should be stressed that social discounting for global warming analysis is unlike that in the choice among government investment projects. When the government is deciding how to allocate the $X billion that the public has appropriated for investment in a given year, it must apply the opportunity cost of alternative public investment projects (e.g. dams, roads, education), and these rates are likely to be in the range of 5 to 7 per cent (and perhaps 10 per cent in developing countries). In the case of greenhouse policy, the choice is not, in the first instance, between investment A and investment B. Instead, the policy will extract future income from the economy at large, so that it is withdrawing resources from future consumption as well as investment. The approach suggested here is that the respective GNP shares are the best guide to the prospective sourcing of the resources extracted as between consumption and investment.

As for the consumption discount rate, there is nothing sacrosanct about the logarithmic utility function, and rates other than the expected growth of per capita income might be appropriate. For estimation purposes, one approach might be to conduct surveys ("contingent evaluation") to determine the discount rate people implicitly apply in thinking about tradeoffs between their own consumption and that of their descendants[7].

A weighted average of the consumption and investment discount rates (for example, 0.85 x 0.75 per cent + 0.15 x 7 per cent) yields discount rates in the 1-2 per cent annual range. On this basis, the "low-intermediate" option in Table 1 would appear the most appropriate, although sensitivity analysis might appropriately include a zero discount rate and a higher rate[8]. Whatever rate is chosen will dramatically influence the calculation, because of the long time horizon.

Other dimensions of analysis

As indicated in Table 1, the analysis will depend in addition on the baseline chosen for warming; on the severity of the policy scenario selected; on the degree of risk aversion; and on whether equity weighting is applied. The baseline for long-term emissions is subject to wide variation (e.g. Nordhaus and Yohe, 1983). On the abatement cost side, the most recent projections by Jorgensen and Wilcoxen (1990) appear to be substantially lower than those of Manne and Richels (1990), in considerable measure because of a lower assumed baseline. The extent of warming varies not only with the emissions baseline, but with the range of GCM projections chosen. With $\lambda = 2.5°C$, the IPCC has set the lower bound of warming under the carbon dioxide equivalent doubling at 1.5°C, and the upper bound at 4.5°C. A corresponding proportionate range would apply to the higher long-term estimates.

An issue for the estimation is whether specific attention should be given to risk aversion. A prior question is whether policy-makers are, or should be, risk averse. If so, the analysis could incorporate non-linear weights on the size of the damage. More simply, the analysis could give greater weight to the upper-bound damage estimates than to the lower-bound estimates in obtaining a risk-weighted expected value of damage. However, there are some concerns that are unlikely to be captured by these approaches. Some analysts have formulated the policy issue in terms of avoiding some vaguely defined "catastrophe", of essentially unestimated magnitude, that is implicitly well beyond even the damage associated with the main estimates for the upper-bound warming ranges. Some form of runaway positive feedback (release of carbon from vast resevoirs in methane clathrates, for example) might be a legitimate motivation for the "catastrophic" approach. However, the "catastrophic" approach simply does not lend itself to analysis on the benefits side, because there is no basis for estimating prospective magnitudes of the catastrophe. Instead, this approach collapses quickly to a truncated analysis that focuses solely on the abatement cost side and asks merely (*a*) what is the least-cost form of abatement?; and (*b*) when is the efficient time to undertake it in view of the prospective delay before scientists and policy-makers "know" whether the catastrophe is or is not a real threat (Manne and Richels, 1990*b*)?

Finally, there is the question of whether the cost-benefit analysis should incorporate equity weights. This issue would arise forcefully in the estimation of benefits from avoiding loss of life. Under "normal lifetime-earnings" approaches, hundreds of deaths in Bangladesh would equate with only a few in Texas. Equity weighting would presumably make the exchange rate much closer to unity. In a less extreme form, the same issue comes up in aggregating the benefits of crop damage avoidance for the United States, on the one hand, and sub-Saharan Africa, on the other (for example). The latter are likely to be much larger relative to income than the former, but in absolute size, the reverse is true. Similarly, losses in poor countries are far more likely to push populations below a poverty line than are damages in industrial countries. Equity weighting may alter the relative importance of different effects, even within the same country. Losses in agriculture, and thus, a basic (income inelastic) good, food, might warrant a heavier equity weight than losses in the areas of leisure activities – typically luxury (income elastic) services, such as skiing.

The equity issue also arises pervasively on the cost side. The most fundamental problem in greenhouse policy may well be the issue of how developing countries can be expected to curtail their emissions when their per capita levels are far lower than in the already-developed countries. Yet, in the absence of action the emissions of developing countries are likely to grow to by far the bulk of global totals.

Overview

Let superscript "u" represent the upperbound damage estimates (e.g. with $\lambda^u = 4.5$, and with the higher baseline for emissions); "l", the lower-bound estimates; and "c" the central estimates. Let "w" denote the risk weighting ($w^l + w^c + w^u = 1$). Let λ_j be the equity weight attributed to geographical area j. Let ρ be the time discount rate. Then, with the "prime" denoting benefit estimation taking account of the adaptation alternative (equation [6]), the discounted present value of global benefits from abatement over the long-term amounts to:

[7] $$PVGB = \sum_i \sum_j \sum_t \lambda_j \{w^l \, b'(i,j)^l_{p,t} + w^c \, b'(i,j)^c_{p,t} + w^u b'(i,j)^u_{p,t}\}/(1 + \rho)^t$$

ESTIMATING GREENHOUSE DAMAGE:
A SURVEY

Agriculture

General considerations

Global warming is likely to damage agriculture in many areas, but aid it in others. The principal damage will arise from heat stress and decreased soil moisture. Warmer temperatures cause the growing cycle of the plant to accelerate, allowing less time for plant development before maturity. Thus, even under irrigation, corn yields are estimated to decline by as much as 50 per cent in the Great Lakes region with GFDL temperature increases (Ritchie, Baer and Chou, pp. 1-1). Moreover, the continental interiors are expected to become drier with global warming. The principal offsets would be longer growing seasons in cold Northern climates, and the fertilization effect of greater atmospheric carbon dioxide. Kane, Reilly, and Buklin (1989) provide the following overview:

> The warming and increased precipitation in the northern latitudes [near 70° N] would improve the agricultural prospects in Canada, northern Europe, and the [former] USSR. ... The increases in temperature and drying of interiors of continents in the mid-latitudes [near 40° N and S] is expected to lead to reductions in agricultural productivity in the United States and western Europe. Projected coastal inundation of rice growing regions in parts of Southeast Asia, such as Bangladesh, combined with the projected movement of the Asian monsoon away from the Indian subcontinent could lead to reduced agricultural production.

On the basis of EPA and IIASA/UNEP studies, the three authors estimate global warming from carbon dioxide-equivalent-doubling would *reduce* average yields by 20 per cent in the United States and EC, and 18 per cent in Canada, but *increase* yields by 15 per cent in Northern Europe and Australia, 10 per cent in the former USSR, and 4 per cent in Japan.

Despite such estimates, it is an emerging stylized fact that the damage to agriculture from global warming will be moderate:

> "because most farmers are not dumb, but rather are accustomed to continually adapting ... [and] the projected impacts of climate are small in comparison with impacts associated with government subsidies, international trade, and other economic factors or ... technological progress" (Ausubel, 1990).

In his review of the benefits of abatement, Nordhaus (1990) places greenhouse damage (double CO_2-equivalent) to US agriculture at ± $12 billion annually (1981 prices), and thus a mean value of zero, based on EPA estimates (discussed below).

For its part, the IPCC concludes:

Negative impacts could be felt at the regional level as a result of changes in weather, diseases, pests and weeds associated with climate change ... There may be severe effects in some regions, particularly in regions of high present-day vulnerability that are least able to adjust technologically to such effects. There is a possibility that potential productivity of high and mid-latitudes may increase because of a prolonged growing season, but it is not likely to open up large new areas for production, and will be largely confined to the Northern Hemisphere. On balance, the evidence is that in the face of estimated changes of climate, food production at the global level can be maintained at essentially the same level as would have occurred without climate change; but the cost of achieving this is unclear. Nonetheless, climate changes may intensify difficulties in coping with rapid population growth (IPCC, 1990c, p. 12).

Thus, the IPCC essentially adopts the "benign" interpretation, but with caution, and a recognition of likely damage to especially-vulnerable regions. There are several reasons to suspect that the "benign" view is too optimistic. Most important, it fails to consider effects over the very-long term. As analyzed below, with eventual global warming on the order of 10°C, agricultural damage would be likely to rise non-linearly. However, even for the standard benchmark of CO_2-equivalent-doubling, there are grounds for skepticism. First, recent work on drought has concluded that the incidence of severe droughts would be greatly multiplied (Rind *et al.*, 1990). Second, the moderate impact projections usually are premised on an important boost from carbon dioxide fertilization. Yet, it is unclear that this phenomenon will prove as effective in the field (i.e., in the farms around the world) as in the laboratory. Moreover, the amount of carbon dioxide buildup is considerably less than two-fold for the doubling of carbon dioxide equivalent of all trace gases.

Consider drought Rind *et al.* (1990) use GISS GCM results to calculate that the incidence of severe droughts (measured by the Palmer index), that currently occur only 5 per cent of the time, would rise to a 50 per cent frequency by the 2050s. Their finding is based on the difference between precipitation and potential evapotranspiration (E_p). They find that:

"E_p increases most where the temperature is highest, at low- to mid-latitudes, while precipitation increases most where the air is coolest and easiest to saturate by the additional moisture, at higher latitudes... Higher temperatures increase the atmospheric water-holding capacity ... of the order of 33 per cent ... and thus the ability of the atmosphere to draw moisture from the surface" (p. 9983).

The GISS GCM calculates E_p as a function of the difference between the specific humidity at ground level and 30 m height (as well as wind velocity and turbulence). E_p rises strongly with ground temperature. Actual evaporation equals potential multiplied by an efficiency factor ($ET = \beta E_p$). The authors argue that the GCMs tend to understate the efficiency factor (at estimates of $\beta \approx 0.2$, versus the observed US range of 0.6 to 0.9), in part because of absence of a vegetative canopy in the models. The overall thrust of the Rind *et al.* analysis is that the GCMs understate the decline in soil moisture in projections of warmer climates. The implication is that agricultural estimates based on the GCM projections may understate the damage from global warming.

Similarly, Gleick (1987) uses GCM climate projections and a water-balance model for the Sacramento basin in California to calculate that summer soil moisture would fall by 8 to 44 per cent under carbon dioxide equivalent-doubling.

The crop modelers acknowledge that the drought issue must be sorted out, but they note that there are offsetting factors, and they maintain that the crop models already incorporate soil moisture reasonably well. For example, plant phenology (life cycle) would be speeded up under a warmer climate. The most severe drought estimates by Rind *et al.* are for June-August, and in a warmer climate crop growth might be completed earlier, in part through adaptive response of earlier planting (Rosenzweig, personal communication). Nonetheless, the drought studies are sobering. In the 1988 US drought alone, losses (primarily agricultural) amounted to $40 billion (Wilhite, 1989).

The second area that may warrant skepticism is the heavy reliance on carbon-dioxide fertilization in the "benign" assessments. Photosynthesis is the transformation of carbon dioxide and water into plant matter (carbohydrates) and oxygen[9]. The so-called "C3" crops (wheat, rice, soybeans, fine grains, legumes, root crops, most trees) have relatively inefficient photosynthesis (because of loss from induced respiration), and benefit substantially from increased exposure to carbon dioxide. The "C4" crops (maize, millet, sorghum, sugar cane) are more efficient and obtain relatively smaller photosynthesis gains from increased CO_2. In addition, both types experience stomatal (pore) closure from increased CO_2, with the result that less water is lost to transpiration and the plants become more efficient in water use. The overall result is that a doubling of carbon dioxide (from 330 to 660 ppm) raises yields in greenhouse experiments by an average 34 per cent for C_3, and 14 per cent for C_4 crops (Schneider and Rosenberg, 1989, pp. 27-28).

However, these effects have not been documented under normal open-field farm conditions. Efforts to detect the 1 to 5 per cent rise in yields that should already have occurred with the rise in carbon dioxide concentration over the past 100 years (e.g. by analyzing control yields of long-established cultivars) have been unable to find a statistically-significant increase against the background noise (Rosenberg and Rosenzweig; unpublished, by communication). The EPA suggests that:

"...studies have examined the interactive effects of CO_2, water, nutrients, light, temperature, pollutants, and sensitivity to daylight on photosynthesis and transpiration ... [or] growth and development of the whole plant. Therefore, considerable uncertainty exists concerning the extent to which the beneficial effects of increasing CO_2 will be seen in crops growing in the field under normal farming conditions with climate change" (EPA, 1989a, p. 110).

Some experts thus emphasize caution on whether open-field carbon fertilization (under changed temperature and rainfall conditions) would be as substantial as the laboratory experiments suggest (Parry, 1990, pp. 40-41; Evans *et al.*, 1991, p. 37). Others stress the large body of evidence supporting at least the laboratory results (Kimball and Rosenberg, 1990) and note that some recent open-field experiments are beginning to confirm those findings[10].

A different type of limitation of assumed CO_2 fertilization is that it is expected to increase plant mass relative to nutritional content. The nitrogen content of plants would decrease as the carbon content increases, "implying reduced protein levels and reduced nutritional levels for livestock and humans'' (Parry, 1990, p. 41). The ''effective yields'' in nutritional terms could thus rise by less than weight yields.

In addition, CO_2 fertilization tapers off relatively quickly with rising atmospheric concentration. For maize, there is no further enhancement at concentrations above 550 ppm. For wheat, the rate of photosynthesis rises from about 50 mg CO_2 dm^{-2}h^{-1} at 350 ppm (approximately current concentration) to 62.5 mg at 550 ppm; the rate rises only to 68 mg at 750 ppm, and rises no further than a 70 mg ceiling for 850 ppm and higher (Parry, 1990, p. 38). As discussed below, the exhaustion of direct carbon dioxide enrichment in the face of above-linear increases in crop damage from heat stress and soil moisture decline mean that agricultural losses under very-long-term warming would be expected to far exceed those for the traditional carbon-doubling benchmark.

Another difficulty with the interpretation of the typical agricultural results is the tendency to mistakenly link double carbon dioxide *content* (and its attendent enrichment) with the *warming* associated with carbon dioxide-*equivalent* doubling. Because of other trace gases, there would be considerably less than double carbon dioxide at double total trace gas equivalent. Specifically, the IPCC projects, in its ''business as usual'' case, that by the year 2025, the radiative forcing from all trace gases would be slightly higher than the equivalent of a doubling of carbon dioxide above preindustrial levels. At that time, atmospheric concentration of carbon dioxide would be enough to generate 2.88 watts per meter squared (wm^{-2}) radiative forcing. Using the IPCC/ Wigley formula (R=6.3 ln c/c$_o$, where ''R'' is radiative forcing, ''c'' is carbon concentration, and ''c$_o$'' = 280 ppm, the preindustrial level), we may calculate that the IPCC expects only 442 ppm carbon dioxide concentration at the equivalent-doubling benchmark[11]. Most of the agricultural studies use 330 ppm as the base level for atmospheric concentration, and 660 ppm for the ''doubled amount'' in calculating yield impacts (EPA, 1989a, p. 100). So, *actual atmospheric carbon concentration is only expected by the IPCC to rise by about one-third the amount typically assumed in the agricultural estimates that incorporate ''carbon dioxide doubling'' enrichment.*

In sum, there are several reasons to be concerned that the present estimates of agricultural effects of global warming are too optimistic, and that even for the modest benchmark of doubling of carbon-dioxide equivalent agricultural damages would be larger, and gains smaller, than in the current ''mainstream'' calculations.

Damage estimates: United States

The EPA presented the following summary estimates of agricultural damage in its 1989 report to Congress. Combined producer and consumer economic effects of global warming at benchmark carbon dioxide equivalent doubling would range from –$5.9 billion annually (GISS model) to –$33.6 billion (GFDL model) at 1982 prices, before taking account of carbon fertilization. After including the effect of carbon fertilization at an increase from 330 ppm to 660 ppm, the estimates were +$10.6 billion using GISS and –$9.7 billion using GFDL (EPA, 1989a, p. 104)[12]. It is apparently these two estimates that formed the basis for Nordhaus' 1990 estimate of ±$12 billion at 1981 prices.

The EPA text, but not its summary table, cautions that the estimates incorporating carbon fertilization may be too optimistic, because they assume *double carbon dioxide (from 330 to 660 ppm),* even though the weather scenarios are for equilibrium double carbon dioxide-*equivalent* warming. It acknowledges that concentration will be only about 450 ppm in 2030 when the doubling-benchmark arrives, and warns (p. 100) that although the "carbon fertilization" variants apply 660 ppm, actual concentration would still be only 555 ppm in 2060, when the equilbrium warming effects of the 2030 are felt (after 3 decades for ocean thermal delay).

Actually, the potential bias is even greater, because the concept under investigation is the equilibrium effect of the 2xCO$_2$-equivalent benchmark. Whereas the 2060 climate in the EPA calculations (GISS Transient Scenario A) is approximately equal in terms of temperature and precipitation to the IPCC's equilibrium climate for doubling CO$_2$-equivalent (as might be expected from the fact that the two diverge in time by about the amount required for thermal ocean lag), the transient EPA benchmark has considerably higher carbon concentration than the IPCC's equilibrium "doubling". It would be inappropriate to count even 555 ppm carbon as available for fertilization at the equivalent-doubling benchmark, because if carbon dioxide trends continue along the path that would take concentration to 555 ppm by 2060, then the warming commitment would have been increased substantially beyond the equivalent-doubling mark. Essentially, to combine even 555 ppm carbon concentration with the GCM warming projections is to mix a transient concept (concentration) with an equilibrium concept (equivalent-doubling).

A proper interpretation of the EPA estimates thus requires a careful weighting of the two alternative sets: with and without carbon fertilization. For measurement of the equilibrium effects of carbon dioxide-equivalent doubling, the proper amount of carbon fertilization to accept corresponds to the IPCC's 440 ppm, not 660, or even 555. On this basis, it is appropriate to give a weight of two-thirds to the estimates "without carbon fertilization" and one-third to the estimates "with fertilization". That is, the actual increase in carbon concentration, from 330 ppm to 440, is only one-third of the way to the 660 ppm used in the carbon-enriched calculation. Even this calculation gives the benefit of the doubt to the carbon fertilization effect because it makes no allowance for the possibility (some would say likelihood) that open field results will be less favorable than those in the laboratory, nor for possible lesser nutritional content per plant mass[13].

It should be recognized that this adjustment by linear interpolation may give different results from those that would be obtained by re-running the crop models with a 2xCO$_2$ climate (temperature, rainfall, soil moisture) and a 440 ppm carbon dioxide concentration. The interreactions of climate and carbon fertilization can be non-linear. Nonetheless, linear interpolation is the best available short-hand method for adjusting the crop estimates so that they represent an equilibrium, rather than a transient, concept. The resulting central estimate for the United States is that benchmark CO$_2$-equivalent doubling warming would impose total (consumer and producer) damages of $13 billion annually at 1982 prices[14], or $17.5 billion annually at 1990 prices.

An alternative central estimate based on drought incidence might be formulated as follows. Suppose the Rind et al. estimate of an increase in the incidence of severe drought from 5 per cent to 50 per cent is accepted. Suppose the $40 billion figure for the 1988 drought is applied as the cost of severe drought. On this basis, the expected annual damage from benchmark warming would be: [50-5 per cent] x $40 billion = $18 billion annually. This alternative approach gives an almost identical estimate to that of the appropriately-adjusted EPA figure, although the concept is very different, because the damage under the drought-probability approach is an expected average, reflecting good and bad years, whereas the EPA figure is for a median year.

29

In another important recent study, Resources for the Future has attempted to do an in-depth calculation for the Missouri-Iowa-Nebraska-Kanasas (MINK) area (Rosenberg and Crosson, 1990). This study is unique in that it seeks not only to incorporate carbon dioxide fertilization, but also to take account of farmer adaptation (earlier planting, use of longer season varieties, changes in tillage to conserve water). The study uses the actual climate conditions of the 1930s as an analog for a 2030s climate.

The MINK results find that if climate change alone is considered, warming by 2030 cuts agricultural production in the area by 17.1 per cent. If carbon fertilization (specified at 100 ppm, from 350 today to 450) is incorporated, the production loss is only 8.4 per cent. With on-farm adaptations, but without carbon fertilization, the production loss is 12.1 per cent. With on-farm adaptations and carbon fertilization, production declines by only 3.3 per cent[15]. Equilibrium long-term losses in the region's forests are 20 to 50 per cent even with carbon fertilization, and streamflows decline by 28 per cent in the Missouri and Upper Mississippi basins, and 7 per cent in the Arkansas basin, suggesting that forestry and water supply may be areas of relatively greater damage.

At first glance, the MINK results (among the most careful of all the agricultural studies) might seem to imply that with carbon fertilization and farmer adaptation taken into account, agricultural damages from global warming would be limited for the United States. A second look suggests otherwise. The major reason is that the 1930s climate was only 1.1°C warmer than that of the 1951-80 base period (Rosenberg et al., 1990, Table 3.2). In contrast, the warming associated with benchmark CO_2-equivalent doubling for the MINK area would be 3.8°C for the average of three high-resolution GCMs (IPCC, 1990a, Figure 5.4, summer and winter average), or if those three models are downscaled to set the climate sensitivity parameter $\lambda = 2.5$ (the IPCC's best estimate), 2.5°C. Benchmark equilibrium-warming is thus more than twice that assumed by the MINK authors. Moreover, the MINK estimates may be favorably biased by using for carbon dioxide enrichment the concentration (450 ppm) that would be associated with the trace gas equivalent-doubling benchmark, even though the warming considered in the study is less than half that standard.

Considering these factors, and if even a mild degree of non-linearity is accepted for the damages, then one would expect the damage associated with the normal warming benchmark to be at least 3 times the MINK estimate. On this basis, *even with carbon fertilization and farm adaptation included, the MINK estimates imply agricultural losses of about 10 per cent from equilibrium warming of carbon-dioxide-equivalent doubling.*

US agricultural production amounted to an estimated $93 billion in 1990 (CEA, 1991). Although there would tend to be gains from warming in Minnesota, Wisconsin, and the Dakotas, the agricultural losses in the South and West would tend to be considerably larger proportionately than those in the MINK states (EPA, 1989a, p. 101; EPA, 1989b, Vol. 1, Section 5). For the United States as a whole, losses would thus conservatively be on the order of $10 billion annually, reformulating the MINK results for compatibility with the usual warming benchmark. The estimate is thus about two-thirds that obtained here after adjusting the EPA central estimate, even though the MINK study incorporates farm adaptation.

Moreover, explicit treatment of loss of consumer surplus could raise all the estimates further. In a simple model of agricultural supply and demand, it is suggested elsewhere (Cline, 199b) that, as an approximations, the total consumer and producer surplus lost from lower output is at least as large as the yield decrease multiplied by the initial value base. That result attains for the horizontal supply curve. With the geometrically-rising supply curve implied by the Ricardian concept of declining fertility of marginal land, combined producer and consumer surplus losses would tend to be higher, perhaps much higher.

To recapitulate, for the United States the agricultural damages from benchmark CO_2-equivalent global warming would amount to some $18 billion annually (both the adjusted EPA basis and the Rind *et al.* drought approach), with a floor of perhaps $10 billion (MINK adjusted).

International yield estimates

For other countries, estimates tend to be in the form of calculated changes in yields rather than value-based estimates. However, by combining yield impacts with estimated value base, it is possible to obtain a minimum estimate of producer and consumer losses (under the proposition just stated). For Canada, Parry (1990) estimates that spring wheat yields will fall by 15 to 37 per cent, although winter wheat should improve. Drought frequency would multiply 13-fold in Saskatchewan. Maize and soybean yields would fall in Ontario. The IPCC places spring wheat yield reductions at 18 per cent (IPCC, 1990b, pp. 2-15), based on estimates by Williams and by Smit.

In Europe, a large decline in summer soil moisture is expected for the south, along with a major reduction in agricultural potential (Parry, 1990). In the former USSR, production could improve in Siberia, but soils there are relatively unproductive (Flavin, 1989). Studies in the former Soviet Union suggest that a decrease in soil fertility, an increase in salinity, and increased soil erosion would more than offset prospective warming benefits, and in the Leningrad and Perm areas, yields of winter rye and spring wheat would decline because of premature ripening of crops (Pitovranov, as cited in Parry and Carter, 1989).

For northern Japan, yields of rice, corn, and soybeans had been predicted to rise by some analysts (Yoshino, as cited in Parry, 1990). However, as noted below, preliminary estimates by an international team assembled by EPA find negative yield effects.

The Scandinavian countries are seen as standing to gain the most from global warming effects on agriculture (IPCC, 1990b, pp. 2-16). For China, the IPCC cites mixed results, with a stronger summer monsoon threatening greater flooding in southern China; possible increases in rice, maize, and wheat yields nationally; but maize decreases in the eastern and central regions (pp. 2-19).

Most analysts see central America as being particularly vulnerable to decreases in soil moisture (Parry, 1990; Stevens, 1990). In the Sahel, an increase in evapotranspiration is expected to exceed the increase in precipitation and reduce soil moisture (Parry, 1990). Drier soils are expected to reduce yields in Northwest and West Africa, the Horn of Africa, and South Africa (Stevens, 1990). The Middle East is also considered to be vulnerable.

The Appendix summarizes prospective agricultural effects, as identified by Working Group II of the IPCC. More specific estimates are available from the underlying studies consulted by the IPCC, and from more recent studies. Table 2 presents an overview of the available yield estimates. Most are drawn from the summary provided by Kane, Reilly and Bucklin (1989). For Australia, China, and Japan, the estimates are identified merely as "negative," based on preliminary results of an international team conducting research

Table 2. **Estimates of Yield Changes From Global Warming**

Country	Location	Crop	% Yield change
Australia[1]	–	–	Negative
Belgium	–	Wheat, Spelt	7
Canada	Saskatchewan	Spring Wheat	–18
China[1]	–	Maize	Negative
Denmark	–	Wheat, Spelt	10
EEC	–	–	–20
Finland	Helsinki	Spring Wheat,	10
		Barley, Oats	9, 18
	Oulu	Spring Wheat,	20
		Barley, Oats	14, 13
France	–	Wheat, Spelt	–11
Germany	–	Wheat, Spelt	–5
Iceland	–	Hay	64, 48
Italy	–	Wheat, Spelt	–1
Japan[1]	–	Rice	Negative
Netherlands	–	Wheat, Spelt	1
USSR	Leningrad	Rye	–13
	Cherdyn	Spring Wheat	–3
	Saratov	Spring Wheat	13

1. See Text.
Source: Kane, Reilly, and Bucklin (1989).

under EPA auspices (EPA, 1990). If confirmed, these initial findings will represent important reversals for China and Japan in comparison with the previous IPCC surveys.

Table 3 applies the estimates of Table 2 to the value of the agricultural base in each country in question to obtain an impression of agricultural effects globally[16]. The final two columns apply the central percentage yield change to the initial agricultural GDP base with a multiple ranging from $\alpha\beta = 0.75$ to $\alpha\beta = 1.25$, where α is the ratio of the output reduction after the incorporation of low-cost adaptational changes to the initial yield reduction, and β is the ratio of total change in consumer and producer surplus to volume output reduction at base period prices. As suggested above, β is likely to be substantially greater than unity. It should be noted that the estimates of Table 3 do not take account of trade patterns, and essentially treat all consumer and producer surplus effects as proportionate to the production base, whereas the presence of international trade means that consumer surplus losses would tend to be relatively greater for food importers, and producer surplus effects relatively greater for food exporters. Nonetheless, the estimates provide a first approximation of international effects.

The estimates of Table 3 suggest that, worldwide, the central expectation would be for net agricultural losses from benchmark greenhouse warming; This result is less optimistic than the IPCC suggestion that world agricultural output (or costs) could change by ±10 per cent, depending on climate, soil moisture, and carbon dioxide enrichment effects (IPCC, 1990*b*, pp. 2-24). One reason appears to be the reliance of the IPCC on favorable effects for China in the rough estimate globally.

The central estimate of Table 3, in contrast, is that the adverse effect of $2\times CO_2$-equivalent warming amounts to a loss of some 7 per cent of world agricultural

Table 3. **Rough Estimate of Impact of Global Warming (2XCO$_2$)**
on Agriculture in Major Countries

Country	1988 GDP ($ billion)		Yield change %	Implied value change	
	Total	Agriculture		α β = .75	α β = −1.25
Australia	205	8.2	−10[1]	−0.6	−1.0
Belgium	145	3.0	7	0.2	0.3
Canada	440	17.6	−18	−2.4	−4.0
China	360	120.8	−10[1]	−9.1	−15.2
Denmark	95	4.1	10	0.3	0.5
Finland	90	5.2	15	0.6	1.0
France	900	30.8	−11	−2.5	−4.2
Germany	1 130	16.5	−5	−0.6	−1.0
Iceland	5	2.0[2]	50	0.8	1.3
Italy[2]	765	31.1	−10	−2.3	−3.8
Japan	2 850	65.4	−5[1]	−2.5	−4.2
Netherlands	215	8.5	1	0.1	0.2
Former USSR	1 000[3]	200.0[3]	0	0.0	0.0
USA[4]	4 885	89.8	−20	−13.1	−21.9
Total	12 815	603.0	−6.9	−31.1	−52.0

1. Postulated. See text.
2. Based on Parry, 1990, p. 87.
3. Rough estimates.
4. Based on text estimate of $17.5 billion for 1990.
α = Ratio of net output effect at given product price after adaptation to initial effect.
β = Ratio of total producer and consumer surplus change to net output effect evaluated at base price.
Sources: Table 2, World Bank, World Development Report 1990; IMF – Note: does not take account of importer versus exporter status.

production (or about $40 billion annually for the countries considered, midpoint of the two variants). The loss is largest in both absolute and relative terms in the United States. The large proportionate gains in northern countries operate on a sufficiently small base that they do not compensate for losses in the United States, China, Canada, Italy, and Japan.

Agriculture under very-long-term warming

The crop models used in many of the estimates of Table 2 could be applied to global warming of 10°C for very-long-term warming. If they were, the results would almost certainly be catastrophic reductions in crop yields. Crops typically cannot grow at temperatures above 35°C (wheat, barley, oats, rye) to 45°C (corn, rice, sorghum; OMB/ USDA, 1989). As noted above, the carbon dioxide fertilization effect is exhausted at about 800 ppm (and earlier for maize). Evapotranspiration tends to rise more than linearly with temperature, and thus outstrips precipitation increases, so that soil moisture under very-long-term warming would be likely to be dramatically reduced.

The broad expectation would be that agricultural losses would be more than linear with respect to temperature rise. Suppose the relationship were only mildly non-linear, less than quadratic; for example, $D = k[\Delta_t]^{1.2}$, where "D" is damage, "k" is a constant, and 'Δ_t' is the change in temperature. Then, with $\Delta_t = 10°C$, instead of $2.5°C$, the damage would be $[10/2.5]^{1.2} = 5.3$ times as large. The initial benchmark world agricultural loss of 7 per cent would rise to 37 per cent. Under these conditions, the consumer surplus losses could rise dramatically, as agricultural output became scarce in the face of inelastic demand.

In short, with a central estimate for agricultural losses on the order of $18 billion annually for the United States and $40 billion world-wide for the normal benchmark of $2xCO_2$-equivalent warming ($2.5°C$), the corresponding estimates for very-long-term warming would be in the range of $95 billion and $212 billion, respectively, even before allowing for greater proportionate consumer welfare losses (and without scaling to future economic size).

Forests

Global warming is expected to cause a poleward migration of forests, and a change in their composition. Sedjo and Solomon (1989) use the Holdridge Life Zone classification system, together with GCM projections of changed climate by grid area, to calculate global forest changes from $2xCO_2$ warming. They find that boreal forests would decline by 40 per cent and temperate forests by 1.3 per cent, whereas tropical forests would rise by 12 per cent (biomass; area figures are comparable). The net change would amount to a decline of 3.7 per cent globally in biomass, and 5.8 per cent in area.

However, these estimates are equilibrium measures after enough time has passed for complete migrational adjustment to the new climate. The potential northern range of forest species in the United States is expected to shift northward by 600 to 700 km over the next century from global warming, and the southern boundary could move north by 1 000 km. However, the actual migration pace could be as low as 100 km over the same period because of slow migration capacity (EPA, 1989a, p. 71). The implied result would be a "temporary" decline in forested area for some three centuries or more, because of tree death on the southern boundary in excess of additional growth on the northern boundary, and thus much larger forest loss than after eventual equilibrium is reached.

Studies cited by EPA indicate that over the next 100 years US forests could lose 23 to 54 per cent of standing biomass in the Great Lakes region and 40 per cent in Western forests (EPA, 1989a, pp. 83-84). On this basis, it might be estimated that global warming from $2xCO_2$ could cause a loss of 40 per cent for US forests. Some portion of this loss could be avoided by increased planting to locate the replacement species in the areas of in-migration sooner than they would naturally arrive. The EPA expresses doubts about this response because:

> "... seedlings that would appear to be favored on some northern sites several decades in the future may not survive there now because of constraints imposed by temperatures, day length, or soil conditions" (EPA, 1989a, p. 86).

In 1989, the US logging industry produced $13 billion in gross output (Commerce, 1990, p. 6-2). The wage bill for 92 000 employes amounted to $1.7 billion. If, as in most US industries, the ratio of wages to capital services was at least 2 to 1, total factor inputs cost approximately $2.6 billion. The value of the wood extracted was thus approximately $10 billion. In view of the prospective 40 per cent forest loss, these estimates would imply greenhouse warming damages on the order of $4 billion annually for the United

States, just for the commercial portion of forests. It is unlikely that these losses would be offset by higher cutting rates, as the current rates of extraction already reflect optimum considerations[17].

Neither the EPA nor other authors appear to have placed quantitative estimates on forest loss[18]. The EPA does suggest that, despite doubts about successful anticipatory planting for migrating species, US reforestation efforts would need to double or triple above their current rate of 0.4 per cent of forest area annually at cost of $400 million annually (EPA, 1989a, p. 89). On this basis, some $600 million in additional reforestation costs could be expected. If the increased reforestation limited the potential forest loss by one-third, the net annual loss would amount to [2/3 x $4 billion + $600 million] = $3.3 billion[19].

The EPA study does not incorporate possible (beneficial) effects of carbon dioxide fertilisation on forests. Instead, its authors cite the likely limiting factors of water, nutrients, and light under canopy, as reasons for "the belief that CO_2 enrichment may not significantly affect forest productivity" (EPA, 1989, p. 81). The outcome might be different for intensively-managed plantation systems, where water and nutrient constraints might be controlled to permit greater exploitation of potential carbon fertilisation effects. However, for at least the United States, the extent of forest under plantation management is a small fraction of total forest area[20]. On the side of "negative influences", the EPA study also omits consideration of "possible increases in fires, pests, disease outbreaks, wind damage, and air pollution" (EPA, 1989, p. 71)[21].

In the very-long-term, two factors would work in opposite directions on the value of forest losses. Eventual completion of migration would potentially reverse the major loss associated with the difference between potential and actual migration rates. However, in the absence of policy changes, warming could be expected to increase without interruption over the next 250-300 years, so there would be a continual lag of forest migration behind that of potential boundaries. Thus, little could be expected from achievement of new equilibrium (because, over the time frame, none would be attained). At the same time, much higher temperatures and drier conditions would be likely to cause additional damage to the forests. A plausible conclusion is that, under "business as usual" conditions, there would be no reduction in the annual losses even over the very-long-term, and instead there would be further increases. In the absence of more formal analysis, something on the order of twice the loss associated with 2xCO2-equivalent might be appropriate for the very-long-term loss.

Internationally, the shift toward tropical (away from boreal and temperate) forests suggests that most developing countries would experience relatively less forest losses, and mid-to high-latitude countries, relatively greater ones. In this important instance, the pattern of relative gains by such northern areas as Scandinavia from global warming, would thus be likely to be reversed[22].

Research in progress for the EPA finds that, globally, approximately one-half of all land area would show a shift in vegetation life zone as the consequence of benchmark carbon dioxide equivalent doubling. Tundra and desert areas would decline by approximately 40 per cent and 20 per cent, respectively, whereas dry forest would increase by about one-third and wet forest would show little change (positive for some GCMs, negative in others). (Smith, Shugart, and Halpin, in EPA, 1990.) Although the expansion of forest at the expense of desert would be felicitous, the estimates are equilibrium changes, and thus major forest losses over a period of perhaps 2 to 3 centuries would still mean damages over the relevant horizon.

Species loss

Analyses of the extent of species loss from global warming are sparse, and, so far, quantification within a corresponding economic valuation remains non-existent. The EPA (1989a) has noted in general terms the risk of increased species extinction, because of changes in habitat and predator/prey relationships, as well as physiological changes. It indicates major reductions in population, but not species loss, for shellfish, fish, and waterfowl as sea levels rise, causing saltwater intrusion into wetlands. The study cites estimates that human activity is already reducing the some 10 million species by a rate at least 1 000 times as fast as would occur from natural forces (p. 152). It cites the poleward migration of forests as a major reason to expect stress on species, especially in view of natural and manmade barriers to animal migration. One interesting response might be government purchase of land to establish migratory corridors (Peters, as cited in Goklany (1989).

Some experts suggest extremely severe species loss from global warming. McLean (1989) notes that the last comparable warming, by 5°C at the end of the last ice age (10 000 - 12 000 BP) was accompanied by massive extinction of the earth's large mammals. He cites embryo death as the likely reason. Under heat stress, blood flow is diverted from the uterus to peripheral tissues to dissipate the heat. Heat stress can produce proportional dwarfs and skeletal abnormalities. Because the large mammals have lesser skin surface relative to body mass, they are the most vulnerable. Moreover, the earth today is already on a relatively high thermal plateau. McLean reaches the provocative conclusion that global warming of 4°C by 2050 (a particular GISS scenario) "could trigger general collapse of mammalian faunas of the middle latitudes."

Economists have identified "use", "option", and "existence" as the three main types of "species values" for economic purposes. Direct "use" value is the most obvious, and would include for example the use of certain species in the production of medicines. "Option" value refers to the economic value of preserving a species to retain the option that it may be of economic use in the future. "Existence" value is a benefit inherent in the existence of the species, like the value society places on the existence of a particular great work of art. Much of the concern with biological diversity seems to derive from existence value (Sober, as cited in Batie and Shugart, 1989).

The three types of value are much more difficult to actually estimate than they are to conceptualize, in part because some 70 per cent of today's species remain uncatalogued. However, society is already undertaking certain costs that may provide a rough guide to the value of species preservation. As the most recent major example, the US government has declared that 11.6 million acres of timberland in the Northwest will be put off-limits to logging in order to preserve the habitat of the spotted owl (*Wall Street Journal*, 29 April 1991). This area corresponds to 1.6 per cent of the total forest land in the United States (738 million acres; EPA, 1989a, p. 72). As developed above, a working estimate for the commercial value of forest resources is $10 billion annually. Therefore, the land set aside for the spotted owl represents an opportunity cost of $160 million annually.

Many species are already endangered or threatened in the United States, including the bald eagle, the grizzley bear, the Florida panther, the whooping crane, and numerous other fauna and flora (Council on Environmental Quality, 1991, p. 138). Suppose that the incremental endangerment of these and all other species from benchmark $2xCO_2$ warming amounts to a conservative 50 times the present endangerment of the spotted owl alone, measured in some intrinsic sense (whereby the simple number of species is weighted by some concept of relative importance). Suppose also that the public's marginal utility of species preservation diminishes as the importance-weighted number of endangered species saved increases (i.e. species preservation is a "normal" service). Then the public might be prepared to pay at least 25 times as much toward species preservation under the species-threatening conditions of $2xCO_2$ as for the present case of the spotted owl. If so, the annual value of the prospective species damage from benchmark warming would amount to 25 x $160 million = $4 billion. As noted previously, the assumptions here are probably far too conservative. For very-long-term warming, with temperature increases four times as great, a simple linear extrapolation would serve to take account of the combined effects of non-linear physical impact and declining marginal utility.

Before more satisfactory benefits estimates for avoidance of species loss can be made, it will be necessary to obtain a better idea of the prospective extent of losses, ideally with numerous concrete illustrations. With such estimates in hand, it might be possible to use the "contingent evaluation" survey technique for evaluating a public good (e.g. Brookshire *et al.*, 1982).

Sea Level Rise

The Intergovernmental Panel on Climate Change has estimated under "business as usual", actual global warming could amount to 4.2°C by the year 2100 (and the warming "commitment" to even more). The group's central estimate for the corresponding sea level rise is 66 cm (IPCC, 1990*a*, Figure 9.6).

US estimates

James Titus of the US Environmental Protection Agency and his coauthors note that this estimate was lower than those of most previous studies, and was based on the assumption that the Antarctic would accumulate more ice rather than be a source of melting over the next century – a view not universally accepted by glaciologists (Titus *et al.*, 1991). The EPA thus continues to use one meter sea level increase by 2100 as its central estimate.

It should be noted that a point estimate for the year 2100 is not an equilibrium estimate, because the melting continues over lengthy periods. Titus *et al.* note that in the last interglacial period 100 000 years ago, when temperatures were 1°c warmer, the sea level was approximately 6 meters higher than it is today. The vast disparity between 6 metres for one degree warming, and 1 metre for 4 degrees, suggests that the ultimate sea level increase would be much higher than the point estimate at 2100. Unfortunately, neither the IPCC nor the EPA appear to have specified the very-long-term sea level rise associated with a specific equilibrium temperature change. Ideally, one would identify the time path of sea level over, say, the next 300 years as the consequence of *a*) no warming, *b*) benchmark 2xCO$_2$ warming (λ = 2.5°c), and *c*) very-long-term warming (e.g. 10°c).

In its 1989 report to Congress, the EPA estimated that, for the United States, a 1 meter rise in sea level by the year 2100 would require $73 billion to $111 billion cumulative capital costs to protect developed areas through the building of bulkheads and levees, pumping sand, and raising barrier islands (EPA, 1989*a*, p. 123). It is also estimated that there would be a loss of dryland of 4 000 to 9 000 square miles.

EPA recently updated its estimates to incorporate the economic value of the land that would be lost. Morgenstern (1990) indicates that the revised estimates place the economic losses associated with a one meter sea level rise at $10.6 billion annually[23].

Titus *et al.* summarize the damages as follows:

"A rise in sea level would inundate wetlands and lowlands, accelerate coastal erosion, exacerbate coastal flooding, threaten coastal structures, raise water tables, and increase the salinity of rivers, bays, and aquifers. ... Coastal marshes and swamps ... collect sediment and produce peat upon which they can build [so that] most wetlands have been able to keep pace with the past rate of sea level rise [but would be unable to do so] if sea level rose too rapidly. ... Moreover, [where] people have built bulkheads just above the marsh ... the wetlands would be squeezed between the estuary and the bulkhead ... Such a loss would reduce available habitat for birds and juvenile fish ... [For barrier islands] Typically ... the bay side is less than a meter above high water. Thus, even a one meter rise in sea level would threaten much of this valuable land with inundation. ... a one metre rise in sea level would generally cause beaches to erode 50-100 meters from the Northeast to Maryland; 200 meters along the Carolinas; 100-1 000 meters along the Florida coast; and 200-400 meters along the California coast ... [yet]most US recreational beaches are less than 30 metres ... wide at high tide ... [A] one meter rise in sea level would ... enable a 15-year storm to flood many areas that today are only flooded by a 100-year storm ... [A] rise in sea level would enable saltwater to penetrate farther inland and upstream in rivers, bays, wetlands, and aquifers, which would be harmful to some aquatic plants and animals, and would threaten human uses of water."

The authors estimate that, if the shoreline retreats naturally, a one-metre sea level rise would inundate 7 700 square miles of dry land (an area the size of Massachusetts). Seventy per cent of losses would be in the southeast (Florida, Louisiana, North Carolina). The eastern shores of the Cheasapeake and Delaware bays would also lose sizeable areas.

The updated EPA analysis contained in Titus *et al.* reaches the following costs of a one metre sea level rise for the United States, if developed areas are protected. Dry land amounting to 6 650 square miles would be lost, as would 49 per cent of today's wetlands. The value of land lost would be $17-128 billion for wetlands, $21-71 billion for undeveloped dryland, and $14-48 billion for land for dikes. Coastal defenses would cost $27-146 billion for sand, $62-170 billion to elevate structures, and $11-33 billion to construct dikes. The total costs would reach $270-475 billion. The amount would be $128-232 billion for a 50 cm rise, and $576-880 million for a rise of two metres (Titus *et al.*, 1991, Table 9).

The central estimate from the updated EPA estimates is thus a capital cost of $370 billion for a one metre rise in sea levels. Morgenstern's translation of this cost into an annual cost of $10.6 billion implies a discount rate of 2.9 per cent. There are important conceptual questions in obtaining the annualized damage value. Essentially, the costs of sea level rise divide into two types: capital costs of protective constructions, and the recurrent annual cost of foregone land services. The capital costs would be spread over a 100-year period. The land "rent" values would also phase in over the same period. But, because the rental opportunity cost is likely to be set at a higher market interest rate than that appropriate for evaluation of the capital costs, the two components should probably be treated individually.

Of the $370 billion total, $224 billion is capital cost for sand, structure elevation, and dike construction. An appropriate way to evaluate this portion is to consider the present value of a 100 year stream of $2.24 billion capital outlay annually. Discounting at 1.5 per cent (the low rate suggested above), the average annual capital outlay for construction amounts to $1.2 billion. Essentially, the entire capital outlay is not made at the outset, so the annual economic cost is considerably smaller than implied by application of even a three per cent interest rate to the principal.

In contrast, the economic costs associated with land loss can be higher. With a base of 13 000 square miles of wetlands, the 49 per cent loss amounts to 6 440 square miles. Titus et al. note that wetland preservation programs typically cost $30 000 per acre. Using a more conservative $10 000 per acre value of wetland, and a "real" land rental rate of 10 per cent, each acre of wetland lost costs $1 000 annually. The 6 440 square mile loss thus generates $4.1 billion in annual losses[24]. For dryland, the median price in US coastal states is $2 000 (Cline, 1990b). Allowing twice the median for greater value near the coast, a conservative valuation would be $4 000 per acre (still lower than the minimum $6 000 in the Titus et al. study). With a (real) land rental opportunity cost of 10 per cent, this land typically contributes $400 annually to economic activity. The annual cost of the 6 650 square miles dryland loss (midpoint) thus amounts to $1.7 billion.

Overall, this alternative approach yields US damage from sea level rise at $7 billion annually. This estimate is of the same order of magnitude as Morgenstern's $10.6 billion, but it takes explicit account of the long phase-in of the capital expenditures. An important implication is that the composition of total sea level rise costs would be much more heavily weighted toward the loss of annual economic services of the land than toward capital costs of coastal defense construction than is implied in the capital values approach of Titus et al. and Morgenstern[25].

Over the very-long-term, damage from sea level rise could be far greater. The Department of Energy has estimated that over 200 to 500 years the West Antarctic Ice Sheet could disintegrate, raising sea levels by 6 metres (studies by Bentley and Hughes, as cited in Titus et al.). Given the long lag from equilibrium warming to equilibrium sea level increase, the one meter figure for 2100 used by the EPA analysts would appear to be likely to be reached even for the equilibrium result of just a 2.5°C global warming. For very-long-term warming, with a horizon of say 250 years (but at 1990 economic scale), the damage would be perhaps four to five times as high.

Other countries

The IPCC (1990b) provides a useful review of the impact of sea level rise internationally. It cites estimates that a one meter rise would inundate 12 to 15 per cent of Egypt's arable land and 17 per cent of Bangladesh. It notes that in flat deltaic areas, such a rise would cause shores to retreat several kilometres, displacing populations. The most vulnerable deltas include the Nile in Egypt; the Ganges in Bangladesh; the Yangtze and

Hwang Ho in China; the Mekong in Vietnam; the Irrawaddy in Burma; the Indus in Pakistan; the Niger in Nigeria; the Parana, Magdalena, Orinoco, and Amazon in South America; the Mississippi in the United States; and the Po in Italy (pp. 6-3, 6-4).

A study prepared by Delft Hydraulics Laboratory for the IPCC estimated the costs of coastal protection shown in Table 4. These costs do not include the value of lost land (although they may overstate dike construction and other defense costs, because they assume that any area with a population density of 10 per km^2 or more would be defended). It is evident from this table that the most severe costs of shoreline defense would be in island nations, with the highest proportionate costs in the Maldives. Thus, it is not surprising that many of these nations have formed the Alliance of Small Island States to press for international agreements limiting emissions (Stevens, 1991). Also according to the Delft study, other countries with coastal defense costs of at least 0.5 per cent of GDP would include Argentina, Bangladesh, Mozambique, and Vietnam (*Change,* 1990).

Work in progress in the EPA's international studies program indicates that, in China, a one-metre rise in sea level could displace over 72 million people, in the absence of measures to hold back the sea (Han, Hou, and Wu, in EPA, 1990). The losses would be concentrated in the Lower Liao River Delta, the North China Coastal Plain, East China Coastal Plain, and Pearl River Deltaic Plain. Shanghai and other major and ancient cities would be completely submerged, and over 125 000 km^2 of agricultural land would be lost. Another study in the same series estimates that, for Egypt, a one metre rise would eliminate one-fourth of agricultural land on the Nile delta and displace eight million people (El-Raey, in EPA, 1990). Other nations at high risk include Thailand and Indonesia (Stevens, 1991). Among the developed countries, the United States would appear to be the most affected by sea level rise (and it has the highest absolute costs in the Delft study, in view of its long and highly developed coastline; *Change,* 1990). In Europe, beaches along the Mediterranean Sea (Greece, Yugoslavia, Italy, Turkey) would be adversely affected, along with related tourism. Venice, which has already experienced a four-fold rise in floods over the last century, would be subjected to increased flooding. The low-lying portion of Poland's coast along the Baltic would be seriously threatened, along with a population of nearly half a million people. In Australia, the Great Barrier Reef would be threatened, as would hundreds of deltas, bays, estuaries, and islands (IPCC, 1990*b,* pp. 6-14/6-15).

Otherwise, the costs of sea level rise for Europe would appear to be relatively moderate. Rosenberg *et al.* (1989*a*) point out that:

"Along the coast from France to Denmark sea walls and dikes already stand at least 16 metres above mean sea level ... to reduce the probability of exceedance to less than 1 in 10 000 years. A 1 metre sea level rise will not be particularly difficult to contend with where such structures already exist" (p. 6).

Similarly, the costs for the former Soviet Union would appear relatively limited.

Space cooling and heating

EPA estimates of increased electricity demand for space cooling, net of reduced demand for heating, as the consequence of global warming have already been reviewed elsewhere (Cline, 1990*b*). The 1989 EPA estimates assumed the GISS scenarios, indicat-

Table 4. Survey of One Meter Sea Level Rise and Protection Costs

Countries Ranked by Estimated Costs as a per cent of GNP[1]

	Country/Territory	Annual Costs as per cent of GNP	Length of Low Coast (km)	Beach Length (km)
1.	Maldives	34.33	1	25
2.	Kiribati	18.79	0	0
3.	Tuvalu	14.14	0	0
4.	Tokelau	11.11	0	0
5.	Anguilla	10.31	30	5
6.	Guinea-Bissau	8.15	1 240	0
7.	Turks and Caicos	8.10	80	5
8.	Marshall Islands	7.24	0	0
9.	Cocos (Keeling) Islands	5.82	1	0
10.	Seychelles	5.51	1	25
11.	Falkland Islands	4.75	1	0
12.	French Guiana	2.96	540	0
13.	Belize	2.93	500	0
14.	Papua New Guinea	2.78	6 400	0
15.	Bahamas	2.67	400	200
16.	Liberia	2.66	2 200	0
17.	Gambia	2.64	400	0
18.	Mozambique	2.48	10 015	25
19.	St. Chr. and Nevis	2.33	40	10
20.	Nieu	2.18	6	0
21.	Guyana	2.12	1 040	0
22.	Surinam	1.94	2 800	0
23.	Sierra Leone	1.86	1 835	0
24.	Aruba	1.85	1	15
25.	Pitcairn Island	1.71	0	0
26.	Fiji	1.53	11	25
27.	São Tome and Pr.	1.46	3	5
28.	Nauru	1.25	3	0
29.	British Virgin Islands	1.24	1	10
30.	Tonga	1.14	4	0
31.	Cayman Islands	1.04	1	25
32.	Cook Islands	1.03	0	0
33.	Equatorial Guinea	1.02	7	0
34.	Antigua and Barbuda	1.01	50	10
35.	Sri Lanka	0.89	9 770	30
36.	Togo	0.87	300	20
37.	St. Lucia	0.82	25	10
38.	Burma	0.77	7 470	0
39.	Benin	0.74	485	10
40.	Micronesia, Fed. St.	0.73	0	0
41.	New Zealand	0.70	14 900	100
42.	Palau	0.69	7	0
43.	Grenada	0.67	6	5
44.	Neth. Antilles	0.66	25	30
45.	Senegal	0.65	1 345	0
46.	Ghana	0.64	2 400	20
47.	Somalia	0.62	700	0
48.	Western Samoa	0.59	11	0
49.	Madagascar	0.56	1 190	0
50.	St. Vincent and Gr.	0.55	7	5

1. This Estimate does not represent the total cost of sea level rise; only the cost of erecting shore-protection structures.
Source: IPCC.

ing 1.2°C warming in the United States by 2010, and 3.7°C by 2055 (EPA, 1989a, chapter 10). The Agency estimated that the resulting increase in electricity demand above baseline would amount to $4.5 billion in annual operating costs by 2010, and $53 billion by 2055 (midpoint estimates, 1986 dollars). In addition, there would be cumulative capital cost increases of $36 billion by 2010, and $224 billion by 2055 (midpoints; p. 192). The EPA study identifed the largest electricity demand increases in the south and mid-continent (20 to 30 per cent above 2055 baseline); intermediate increases in California and mid-Atlantic states (10-20 per cent); more moderate increases in the mountain and north-central states (0-10 per cent); and net electricity demand *reductions* in four north-western states and the three northeastern-most states (0 to −10 per cent; p. 193).

As indicated in the methodological section above, it is useful to state damages against base year GNP, and then deal separately with scaling for future years. Against a 1990-scale economy (rather than the future baseline), expected warming would increase electricity operating expenses by $3.6 billion by 2010 and $18.6 billion by 2055; the corresponding increases in cumulative capital costs would be $29 billion and $89 billion (at 1990 prices; Cline 1990b).

Considering thermal ocean lags, the EPA-GISS estimates for 2055 would approximately correspond to equilibrium estimates for $2xCO_2$-equivalent. However, the GISS model yields somewhat higher warming than the IPCC's λ =2.5° (with the GISS λ = 4.2°; IPCC, 1990a, Table 3.2a). Scaling back to the IPCC's standard, adjusted 2055 estimates (at 1990 economic scale and prices) would stand at $11.2 billion for annual operating costs, and $53 billion for cumulative capital costs (i.e. 60 per cent of the amount before adjustment). Dividing the cumulative capital cost by the number of years, and then discounting at 1.5 per cent annually, and adding the annual operating costs, the additional costs of electricity for space cooling from $2xCO_2$-equivalent would amount to $11.7 billion annually[26].

The EPA estimates, even after scaling back to the size of the 1990 economy, indicate significant non-linearity. The operating costs mushroom from $4.5 billion for 1.2°C warming, to $18.6 billion for 3.7°C warming. If we apply the form $D=k[\Delta t]^\gamma$ and estimate from the two "observations" just mentioned, then γ = 1.26. On this basis, in the case of 10°C very-long-term warming with Δ_t four times as large as for $2xCO_2$-equivalent warming, additional US annual space cooling costs (operating expenses) would amount to $11.7 billion x $4^{1.26}$ = $67 billion (at 1990 economic scale and prices).

The EPA electricity estimates are already net of savings on space heating. However, these savings do not include those for oil and natural gas heating. Nordhaus (1990) has estimated savings on heating costs at $1.1 billion annually (1981 prices), based on the assumption that heating costs would decline by 1 per cent. However, he apparently includes electricity savings in the estimate, whereas the EPA electricity demand estimates are already net of savings on heating. Estimates by the US Energy Information Administration place non-electricity heating energy consumption (residential and commercial) at 7.3 quadrillion BTUs for 1990 (EIA, 1990, p. 30). At an average price of $2.60 per million BTUs for natural gas, residual oil, and distillate oil in 1982 dollars (p. 233), or $3.50 at 1990 prices, each quad costs $3.5 billion. Non-electric heating thus amounted to some $25.6 billion in 1990. If it is assumed that benchmark warming would reduce heating costs by an optimistic 5 per cent, the annual savings would amount to $1.3 billion.

Internationally, the IPCC summarizes energy demand effects as follows. For the United States (based on the EPA study), electricity capacity would need to rise 14 to 23 per cent above baseline by 2055. For Japan, 3°C warming would cause a 5 to 10 per cent rise in electricity demand above baseline. In contrast, for Germany, a 1°C warming is calculated to reduce energy consumption (apparently including non-electric) by 13 per cent for older single-family homes, and up to 45 per cent for new homes, while increasing energy consumption for air conditioning by 12 to 38 per cent. The result is a net decrease in energy demand for space heating and cooling by 12 per cent in 2010. Similarly, for the former USSR, a 1°C warming is expected to generate larger savings in heating costs than increases in air conditioning costs. For the developing countries, the international group noted that a smaller percentage of electricity consumption is devoted to residential and commercial use than in the industrial countries (28 per cent versus 55 per cent for Japan, for example). It thus conjectured that incremental electricity needs from global warming would be less than 10 per cent for developing countries (IPCC, 1990*b*, pp. 5-19/5-20).

Human amenity

As suggested elsewhere (Cline, 1989; 1990*b*), the public might be prepared to pay something to avoid the warming to be expected from global climate change. These estimates of very long term warming indicate that with 10°C global warming, for the 66 largest US cities, the number with July average maximum daily temperatures above 90°F would rise from 18 to 62. The number with July average daily maximums above 100°F would rise from 2 to 42.

Mearns *et al.* (1984) use statistical distributions of current temperatures to explore the impact of global warming on extreme temperature events for the United States. In their base case, they increase mean temperature by 3°F (1.7°C), and hold the variance and autocorrelation of daily temperatures constant. Under these assumptions, they calculate that the frequency of heat waves (defined as 5 consecutive days with maximum temperature at least 95°F) would multiply three-fold (estimated for Des Moines, Iowa). Again, there is reason to believe that people would be willing to pay something to avoid a threefold increase in heat waves.

Global warming could reduce disamenity of severe winters in colder areas, providing some offset to the increased disamenity of heat waves and generally hotter summers. One way to approach estimation would be the survey technique (contingent evaluation), with questions formulated to elicit a meaningful evaluation of what the respondent would be prepared to avoid the change (or, for more clement winters, what he/she would be prepared to pay to enjoy the change). It has been suggested elsewhere (Cline, 1990*b*) that the US public might be prepared to pay 0.2 per cent of its income to avoid increased disamenity of severe high temperatures, or $10 billion annually, but this number is a pure guess.

Disamenity damages in other countries would depend on location and income levels. For high latitude countries, the effect would probably be favorable (i.e. negative disamenity damages). For mid- and low-latitude countries, disamenity effects might dominate, and almost certainly would do so under very-long-term warming. Valuation would tend to be higher for higher income countries, even relative to GDP, under the assumption that amenity is an income-elastic service.

43

Life and morbidity

In the United States, an estimated 1.1 million of the 1.7 million deaths annually are due to diseases that are potentially sensitive to weather (cerebrovascular, pulmonary, diabetes, heart, pneumonia and influenza) (EPA, 1989, p. 221). Heat waves, as well as very cold weather, increase the incidence of stroke and heart attacks. Air pollution increases the occurrence of respiratory diseases (emphysema and asthma), and longer, warmer summers are expected to increase the severity of air pollution (as discussed below).

Kalkstein has carried out statistical analysis relating deaths to weather variables (including temperature and humidity). He has then applied changed weather conditions under $2xCO_2$ to calculate the change in number of deaths for 15 major US cities (EPA, 1989a, p. 224). These estimates assume one case with no acclimatization, and another case with full acclimatization, the latter being estimated on the basis of corresponding mortality statistics for a control city with climate conditions comparable to those in the warmer future for the city in question. The estimates "with acclimatization" yield some anomalies (i.e. higher deaths with than without acclimatization to the new climate; or lower deaths after acclimatization than at present). However, the majority of estimates are in the directions expected. Moreover, increased summer deaths substantially exceed decreased winter deaths. For the unweighted sample, summer deaths rise from 1 156 at present, to 7 402 under warming "without acclimatization", and 2 198 "with acclimatization". Winter deaths fall from 243 at present, to 52 "without acclimatization", and to 159 "with acclimatization" (the latter is probably spurious, since, in winter, the population should be able to do at least as well "with" acclimatization as "without").

The Kalkstein estimates may be reinterpreted by enforcing as the post-warming summer deaths the minimum of the following: present level, post-warming "without acclimatization", and post-warming "with acclimatization". Similarly, post-warming winter deaths may be set as the maximum of present, future with, and future "without acclimatization". When this screening is performed, and when the resulting net death rates per 1 000 population of the individual cities are weighted proportionately to city populations, the overall result is that benchmark warming would increase mortality by 0.0397 persons per 1 000 population. Applied to the present US population base, this result indicates that global warming at $2xCO_2$ would cause some 9 800 additional deaths annually.

Life valuation methods for public policy often take as a point of departure the lifetime earnings of an individual, under the assumption that earnings are what the society is willing to pay the individual (and thus the person's "marginal product"). Average non-agricultural wages in the United States amounted to $17 994 in 1990 (CEA, 1991, p. 336). If a working lifespan is set at 45 years and a 1.5 per cent discount rate is applied, the value of average lifetime wages is $595 000. If this rate is applied to the estimated increase in deaths, annual mortality costs from global warming would amount to $5.8 billion for the United States.

The value of damage could be much greater on the basis of some of the higher life-valuation estimates in the literature (Cropper and Oates, 1990). On the basis of the statistical relationship of wages to risk of death by occupation and industry, estimates of the "value of a statistical life" have been placed in the range of $2 million and even up to $6 million. Contingent valuation studies that ask workers what wage differentials

44

would be required for more dangerous work generate results in the same range ($2 million to $3 million). On the other hand, studies that measure actual behavior in hazard avoidance (e.g. the purchase of smoke detectors) place the value of a statistical life at a much lower range of $500 000 to $600 000. Moreover, for environmental risks that are longer term and lower probability of death (e.g. exposure to hazardous waste), the value of a statistical life is even lower ($200 000 to $400 000), in part because the risk is some 20 years or more in the future.

One issue in valuing a life in the range of $2 million to $3 million is that the implied value of the entire US population would be more than 100 times the size of GNP. Since each worker's productive life is only some 45 years, this implied aggregate amount would be at least twice what American society could afford to pay for its own survival. This apparent anomaly can be resolved, however, if it is considered that for events that would only cause the death of a small fraction of the population, people might well be prepared to pay a higher per-death insurance cost than implied by the budget constraint of per capita lifespan GNP, given risk aversion and the opportunity for risk-pooling.

The incidence of increased deaths would tend to be disproportionately among the elderly. It might be argued that the lifetime earnings approach would consequently apply a substantial reduction to this estimate. However, in other areas of health policy, the US public has shown no inclination to value lives of the elderly at any less than those of (for example) those in the 20-30 year age group. Accordingly, no reduction is applied here for the age distribution of the incremental deaths.

A related issue is that increased deaths from warming might primarily be instances in which the date of death is "merely" advanced by a few years. There might be some grounds for reducing the estimate from this standpoint. However, from another perspective the estimate is already low, as it uses the average wage, rather than total income per capita, and implicitly excludes the deceased's human and financial capital income from lifetime valuation. Use of GDP per capita would increase the loss estimate by about one-fourth.

On balance, the rate used here is probably an understatement of the value of a life saved, compared with most of the estimates in the literature. However, there is a more fundamental question about the estimates of global warming effects. Many in the medical profession are reportedly skeptical about the meaning of the statistical estimates relating death to temperature, and tend to doubt that there would be any major mortality consequences from warming; instead, they tend to be more concerned about morbidity effects[27]. It is unclear, for example, whether the statistical tests have adequately controlled for the tendency of older people to live in warmer cities in retirement. In lieu of more concrete analysis of the medical basis for mortality effects, the estimates here may be seen as compensating for possible overstatement of mortality effects by application of a conservative per-life valuation.

Morbidity could also suffer in the area of "vector borne diseases", where the vectors are such carriers as ticks and mosquitoes and the diseases include Rocky Mountain spotted fever, Lyme disease, malaria, and dengue fever. Simulations for the United States on possible increase of malaria under warming are inconclusive (EPA, 1989a, p. 230). However, there would be a general migration of the vectors (carriers) from south to north.

Migration

Those who approach the greenhouse issue primarly from the standpoint of adaptation rather than abatement, often stress migration as a means of response (e.g. Schelling, 1983, pp. 455-6). This raises two types of costs: utility sacrificed by the migrant, and cost imposed on the target host country.

In the abstract, it seems curious indeed that migration costs would be construed implicitly as being minimal, or even negative, by some in the "adaptation school". To state the issue starkly, peoples have often fought wars to avoid being forced to leave their homelands against their will. There is little reason to suppose that when the cause is climate change rather than an invading army, they will instead be indifferent between staying and moving. There is also a question of the ultimate feasibility of the migration response.

As the amount and quality of land "opened up" by warming becomes smaller relative to the land closed off, the migration option becomes progressively less viable. Only so many people will fit onto Iceland.

Long before people move *en masse* toward more favorable regions, however, they will confront political obstacles to cross-border migration. This raises the second type of migration cost: that imposed on, or at least perceived by, the target country. Despite economists' ability to demonstrate theoretically that free migration would maximize world income (though with distributional effects between capital and labor), unselective immigration is typically seen as negative in political terms. Otherwise, there would be no immigration quotas in industrial countries.

For the United States, it is not difficult to envision increased pressures from legal and illegal immigration caused by global warming. Increased hurricane damage and sea-level rise could intensify the push of immigration from the Caribbean. Increased drought could do the same for immigration from Mexico.

The stock and flow of immigrants provides a rough basis for a sense of the costs that might be imposed by warming-induced migration. Approximately 1.8 million illegal immigrants applied for legalization under the 1986 Immigration Reform and Control Act (IRCA), of whom 70 per cent were from Mexico. In the period 1975-80, an estimated 130 000 undocumented aliens from the Western Hemisphere entered the United States each year (Goering, 1990). The annual flow of legal immigrants has been approximately 640 000 (World Almanac, 1991). The base of legal and illegal immigrants is thus approximately 800 000 annually.

Suppose that benchmark warming ($2xCO_2$) increases illegal immigration by 25 per cent, and legal immigration by 10 per cent. Suppose that on a base of 800 000 immigrants, the share of "illegals" is 20 per cent. Then warming would increase annual immigration by about 100 000. Infrastructure spending (education, roads, police, sanitation) in the United States is primarily at the state and local level. State and local government spending in 1989 amounted to $762 billion (CEA, 1991, p. 383), or approximately $3 000 per capita. Suppose that an immigrant only begins to pay taxes that cover his incremental social infrastructure costs after a period of 18 months (for "illegals", the period could be longer). Then each additional immigrant imposes a cost of $4 500 on the United States[28]. On this basis, an extra 100 000 immigrants annually would represent a cost of $450 million as an adverse effect of benchmark warming for the United States. Europe could also experience increased immigration, primarily from Africa.

Hurricane damage

Emanuel (1987) argues that tropical cyclones are:

"... particularly sensitive to sea surface temperature as the latent heat content of air at fixed relative humidity is a strongly exponential function of temperature, approximately doubling for each 10°C increment of temperature above 0°C".

Emanual calculates that for a rise of 3°C in sea surface temperature, there is a 30 to 40 per cent rise in the maximum drop of barometric pressure. Wind velocity rises as the square root of the drop in barometric pressure. Wind pressure against obstructuctions rises with the square of wind speed. Applying a 2.3°C to 4.8°C range of GCM estimates for increased August temperatures in the tropics (under the $2xCO_2$ scenario), Emanual estimates that the destructive potential of hurricanes would rise by 40 to 50 per cent.

Similarly, Hansen et al. (1989) find in $2xCO_2$ simulations with the GISS GCM that moist static energy (the sum of sensible heat, latent heat, and geopotential energy) rises by 10 kJ/kg at the earth's surface and 8 kJ/kg at the upper troposphere (approximately 40 000 feet). Because of the increase in this gradient, and the greater vertical penetration of moist convection (as well as increased evaporation and higher temperatures at low levels), they conclude that global warming would bring "increased intensity" of "both ordinary thunderstorms and mesoscale tropical storms".

Wendland (1977) has provided empirical support for the relationship of hurricanes to ocean surface temperatures. From monthly data for 1971-81, the frequency of hurricanes is closely related to the size of ocean area with temperature over 26.8°C, and the relationship is exponential.

The Emanuel study provides a basis for quantification of prospective damages. Over the last 40 years, US hurricanes have cause an annual average of $1.5 billion in damages at 1989 prices (Cline, 1990a). The most costly storm on record was Hurricane Hugo in 1989 ($7 billion). On the basis of the 50 per cent increase suggested by Emanuel, the increased damage to be expected from benchmark warming would be $750 million annually. Hurricanes also cause loss of life, typically a total of 50 to 100 annually in the United States.

For other countries, especially island states, hurricane damage from global warming could be more severe. In 1970, tides from a cyclone storm killed hundreds of thousands in what is now Bangladesh. Hurricane Fifi in 1974 killed some 5 000 in Central America. Hurricane Gilbert caused $8 billion in damage in Jamaica alone in 1988 (IPCC, 1990b, pp. 5-10).

The exponential relationship of latent heat content of air (at fixed relative humidity) to temperature, and the threshold effect whereby ocean surfaces above 26.8°C become subject to hurricane conditions, suggest that increased hurricane damage from global warming would be more than linear in relation to temperature rise. Thus, damages under very-long-term warming could be far greater proportionately than the corresponding ratio of temperature increase to that under benchmark $2xCO_2$ warming.

Construction sector

It is often assumed that global warming would benefit the construction sector because "lengthening of the construction season" would be "likely to increase productivity" (Nordhaus, 1990). However, the increased incidence of heat waves (discussed above) would seem likely to eliminate some summer construction days. More importantly, although construction is adversely affected by frost, it is also inhibited by rainfall (Jones, 1964; Russo, 1966, as cited in Solomou, 1990). For example, gravel digging has to stop under conditions of heavy rain.

Solomou (1990) has examined data on construction and rainfall in the period 1856-1913 and found a close negative correlation between the two. The GCMs typically predict an increase in global mean precipation by about 8 per cent as the consequence of benchmark $2xCO_2$ warming, and the estimates range as high as 13-15 per cent (IPCC, 1990a, Table 3.2, p. 87). The IPCC also notes adverse effects of rainfall on industry more generally, and cites Parry and Read to the effect that "rainfall is responsible for more delays than any other climatic variable" for UK industry (IPCC, 1990b, pp. 5-34).

More work is required before a judgement can be offered on whether the adverse effects of increased rainfall on construction (and other industries) would be greater or smaller than the beneficial effects of warming on construction (and other activities) during winter months. The principal thrust of the consideration of rainfall is thus to cast doubt on the idea that there would be much benefit for the construction sector, if any, from global warming.

Leisure activities

Studies for Canada cited by the IPCC indicate that the ski industry would lose 40 to 70 per cent of skiable days as the consequence of warming from carbon dioxide doubling, even after taking account of the availability of snow-making machinery (IPCC, 1990b, pp. 5-35). Losses would presumably be larger in relative terms in US ski areas, where the temperature base is already higher. Other outdoor leisure activities would be much less affected. For example, golf might benefit in cold areas, but be adversely affected in warm regions because of high temperatures, and along heavily populated coastal areas, where rainfall would tend to increase.

Tourism could also be affected in areas where coral reefs are attractions. Coral bleaching and coral death have been observed in the Pacific after the 1982-83 El Niño and in the Caribbean after unusually high water temperatures in the summer of 1987 (EPA, 1989a, p. 157; In Depth, 1990). Bleaching results from the expulsion of symbiotic algae under environmental stress. As these algae are the primary source of food for the coral, the result can be coral death. Stress on coral reefs is expected to occur under global warming, in part because of higher temperatures, but also because vertical accretion of reef flats may be unable to keep up with sea level rise.

Placing a dollar value on impacts in leisure activities is difficult. However, as a point of departure, the US ski industry accounted for an estimated $5.6 billion in economic activity in 1988, when there were 53 million skier visits to ski areas (Waters, 1990). If we suppose that global warming of $2\frac{1}{2}°C$ would reduce ski activity by 60 per cent in the United States, and it is assumed that half of the gross loss is offset by the difference between the value of released labor and other variable factors of production on the one hand and the loss of consumer surplus on the other, then the net loss would be on the order of $1.7 billion annually.

Water supply

Warming is expected to put stress on water supply. A smaller share of precipitation in the form of snow, combined with earlier snow-melt, would mean higher runoff in the winter. Runoff in the spring and summer would correspondingly be reduced. In those areas and/or periods where precipitation declines, the combination of higher evaporation (from warmer temperatures) and lower precipitation would reduce soil moisture and water levels and flows. At the same time, the demand for water would tend to rise with warming, because of increased needs for irrigation and for cooling in electric power production (EPA, 1989a, pp. 165-170).

Average precipitation in the United States is 4 200 billion gallons per day (bgd). Of this amount, 2 765 bgd evaporates, 338 is withdrawn for use, and 1 435 goes to surface and groundwater flows. Of the withdrawals, 140 bgd is used for irrigation, 131 for thermoelectric power, 36 for domestic use, and 31 for industry and mining. Each of these sectors returns a portion of the flows, so that the shares of net consumption are different: 76 bgd for irrigation, 4 for electric power, 7 for domestic use, and 5 for industry and mining (EPA, 1989a, p. 166).

Using a water balance model for the Sacramento basin, Gleick (1987) estimates that a 4°C increase in temperature would decrease summer runoff by 55 per cent, even if there were a 10 per cent rise in precipitation. If there were a 10 per cent *decline* in precipitation in the region, the reduction in summer runoff would amount to 65 per cent. Winter runoff would increase in both cases (but would decrease if precipitation fell by 20 per cent).

The EPA estimates that $2xCO_2$ warming would reduce annual water deliveries in California's Central Valley basin by some 200 000 to 400 000 acre-feet, or by 7 to 16 per cent, even as demand for water rises by some 1.4 million acre-feet by as early as 2010 (EPA, 1989a, p. 251). The price of water in California in a normal year is approximately $250 per acre-foot (Stevenson, 1991). In this one instance, then, the costs of increased water scarcity would amount to $75 million annually (300 000 acre-feet x $250).

Nationwide withdrawals of water amount to 0.378 billion acre-feet annually[29]. Assuming $250 per acre foot for domestic and industrial use (59 per cent) and $100 for irrigation (41 per cent), the annual value of water withdrawals is on the order of $70 billion annually. If water availability would decline just 10 per cent with climate change (based on the Gleick and especially the EPA-California estimates), the annual cost would amount to $7 billion.

Inefficient water allocation could make the costs higher. After five years of drought, in early 1991, California authorities cut off irrigation water to a wide range of agricultural producers. The first desalinization plant for residential water supply in the United States is now planned for Catalina Island in Southern California. Other communities are considering turning to salinization plants. Costs are expected to amount to $1 000 per acre foot (Stevenson). If such sources represent the marginal cost of water, the estimate just suggested could substantially understate the value of water lost to climate change.

One way to ease prospective shortages would be to reform pricing practices so that agricultural users pay a price closer to that for residential and commercial use. In the past, politics has prevented this outcome. Moreover, as discussed above in the analysis of agriculture, the demand for irrigation is highly likely to rise from global warming, especially if the projections of sharply higher drought incidence are accurate.

The IPCC reports on various methods for estimating the impact of global warming on water supply. The group stresses that river basin runoff is very sensitive to small variations in climatic conditions, largely because runoff is a residual between precipitation and soil absorption or evaporation, and small changes on any of the underlying variables can cause a much larger than porportionate impact on this residual. For the United States, water basin simulation models show that in a warmer, drier climate ($+2°C$, -10 per cent precipitation), water supply in 18 major water regions (covering the bulk of national supply) would decline by approximately one-third (calculated from IPCC, 1990*b*, Figure 4.2). Under the same conditions, annual river runoff in regions with relatively low precipitation would decline by 70 per cent (pp. 4-7). These estimates suggest that the 10 per cent decline applied above is conservative.

For Canada, several studies based on GCM estimates for global warming indicate that, in the Great Lakes region, runoff to the lakes could decline by 8-11 per cent (for warming of $3.1°C$ to $4.8°C$ and precipitation change of -3 to $+8$ per cent). Further to the north, in the James Bay region, the studies show increased runoff because of much higher precipitation ($+15$ to $+17.5$ per cent); and for the Canadian Prairie (Saskatchewan River), estimates diverge (IPCC, 1990*b*, pp. 4-6).

For Europe, several studies based on water balance models and GCM projections suggest that there would be reduced precipitation and runoff in the south (Spain, Portugal, Greece), possible decreases in runoff in the central region, and significant increases in the north (the UK, the Netherlands, and Belgium; IPCC, 1990*b*, pp. 4-11/12).

For the former USSR, studies in part based on paleoclimatic analogues have suggested that global warming would tend to increase water resources, as a $2°C$ global warming would increase annual runoff by 10-20 per cent on all the large rivers. An important exception is that, in the south of the forest zones of the former European USSR and western Siberia, annual runoff could fall by 80 per cent or more. (IPCC, 1990*b*, pp. 4-15/16).

The IPCC reports that, for Japan, normal GCM projections are especially inadequate because they do not provide information on typhoons. Japanese experts anticipate that with global warming there would tend to be longer periods of drought, interrupted by intense precipitation. Water shortages could become more prevalent.

Detailed studies in New Zealand conclude that, with global warming, there would be a large increase in annual runoff through most of the country, but a large decrease along the eastern shores of the two islands. The studies indicate a sharp increase in flooding (IPCC, 1990*b*, pp. 4-19; Figure 4-5).

The broad picture that emerges is that, as Gleick (1987) suggests, stress on water supply may be one of the most important consequences of global warming. The prospective decrease in supply in many areas (especially the United States) would confront an increase in demand that could pose severe price increases. Rosenberg *et al.* (1989) recall that, as early as the 1960s, there were discussions in the United States of large-scale water transfers from Alaska and northern Canada to the Southwest and High Plains of the United States. In the former Soviet Union, there was even consideration of reversing the northerly flow of some major rivers (until the large economic and environmental costs became more apparent). Water shortages could revive some of these proposals (at least for the United States), with the certain effect of stirring political conflicts over water transfers between regions and among sectors (especially away from agriculture).

If global warming of benchmark 2.5°C would cause problems for water supply, very-long-term warming in the range of 10°C would seem likely to cause much more dramatic difficulties. It is useful to recall the point made by Rind *et al.* (1990): with a warmer atmosphere, more water is in the atmosphere in the form of water vapor and less is left in the surface land, lakes, and rivers (i.e. not all of the water transferred to the atmosphere comes from the oceans). As Rind *et al.* (1990) note:

"... from the Claussius-Clapeyron equation, an incremental increase in temperature produces a greater increase in atmospheric moisture-holding capacity when the temperature is warmer" (p. 9983).

One would thus expect evapotranspiration to rise more than linearly with temperature, causing a non-linear rise in water availability problems.

Non-linearity is confirmed in the simulations of Gleick (1987, p. 146). He finds that summer runoff decreases by 12 per cent when temperature rises by 2°C and precipitation rises by 10 per cent. Summer runoff decreases by 49 per cent when temperature rises by 4°C and precipitation rises by 20 per cent. Thus, a doubling of both the temperature and precipitation increases leads to a quadrupling of the percentage cutback in summer runoff. Yet there is reason to believe that the result would be even worse, because as just suggested, precipitation would be unlikely to rise as much as evaporation (because of the non-linear rise in atmospheric water content).

In sum, a damage estimate on the order of $7 billion annually might be appropriate for the United States for water supply under global warming with $2xCO_2$. For warming four times as large over the very-long-term, damage would be likely to be much more than four times as high. In the formulation $D=k[\Delta t]^\gamma$, one might expect $\gamma = 1.5$ (for example). In that case, the very-long-term damage would amount to $7 billion x $4^{1.5}$ = $56 billion (against 1990 economic scale).

Urban infrastructure

The EPA has examined the impact of global warming on urban infrastructure costs. For coastal cities, sea level rise or more frequent droughts would increase the salinity of coastal aquifers and tidal surface waters, requiring a response where these are the sources of water. New Orleans illustrates the point. In the drought of 1988, when the Mississippi River was far below normal levels, it was necessary to build a temporary 9-meter silt wall to halt the upriver advance of saltwater that threatened the city's water supply. In the Philadelphia-Wilmington-Trenton area, a sea level rise of just 0.3 meteres could require a 12 per cent increase in reservoir capacity to prevent saltwater from advancing past water intakes on the Delaware River. (EPA, 1989*a*, p. 239). In addition, more frequent and intense storms would likely overload existing storm sewer systems. Flooding and release of untreated waste into watercourses from storm and wastewater systems could require new sewer systems.

A study for New York City found that, with benchmark warming, increased water use for cooling large buildings and for lawn watering could increase annual water demand by 5 per cent, while increased evaporation of water in reservoirs could cut supply by 10 to 24 per cent. Saltwater infiltration from rising sea level could place some intakes below the salt line during the summers with mild droughts, reducing supply further. The EPA summary of various studies for New York estimated that for water supply adjustments alone, the city would need to invest an additional $3 billion as the consequence of $2xCO_2$ warming (EPA, 1989a, p. 243).

In Miami, apart from costs for levees to deal with rising sea levels (considered separately above), capital outlays for canal control, drainage, and raising streets (otherwise subject to collapse from water table infiltration of their base, for about one-third of streets) would amount to some $580 million (p. 241). In contrast, for Cleveland, the savings on snow and ice control and municipal heating would offset higher air-conditioning costs and additional dredging and water supply costs would be negligible (p. 242).

Urban infrastructure investments, primarily those related to adjustments in water supply, sewer, and drainage systems, might amount to something on the order of $10 billion, based on the New York and Miami estimates. Divided over 75 years and discounted, a $10 billion capital cost amounts to approximately $100 million annually[30].

Pollution

The benefits of action to reduce global warming would include two types of gains that stem from the interrelationship of global warming to more traditional problems of pollution. First, there would be a direct "damage avoidance", similar conceptually to the avoidance of damage to agriculture (for example). This effect would arise because a warmer climate could aggravate the problem of air pollution, as discussed below. Second, there would be a spillover benefit from the action undertaken to reduce global warming, to the extent that the action involved a cutback in the burning of fossil fuels that presently contribute to air pollution. For example, if the use of coal is reduced as the consequence of policy action on global warming, there is a spillover benefit in the form of reduced pollution associated with coal burning (e.g. emissions of sulfur dioxide and resulting acid rain). The discussion that follows will examine just the first of these two concepts, but the second is also important.

A warmer climate would aggravate urban pollution. One piece of evidence is that, in the hot summer of 1988, the extended stagnation periods and high temperatures caused 76 cities in the United States to exceed the national ambient air quality standard (NAAQS) for low-level (tropospheric) ozone pollution by 25 per cent or more (EPA, 1989a, p. 200).

Air pollution involves total suspended particulates (TSP), sulfur dioxide, carbon monoxide, nitrous oxide, ozone, and lead. Of these, the most severe problem in terms of the number of persons living in areas with air quality indexes that exceed (violate) the NAAQS is that of ozone (75 million in 1986). TSP and CO also remain leading problems (41 million each; p. 202). Ozone has proven the most difficult to reduce. Its concentrations declined only 13 per cent from 1979 to 1986, compared with reductions of 23 per cent for TSP and 37 per cent for SO_2 (pp. 201-202).

Numerous studies confirm that ozone concentrations rise with temperature (see e.g. IPCC, 1990b, pp. 5-46/47). Models estimated by Morris indicate that a 4°C rise in temperature (e.g. approximately $2xCO_2$ benchmark for the United States) would increase the number of people-hours of exposure to ozone in excess of the NAAQA ceiling of 120 ppb from 661 000 to 2.1 million in Central California, and from 29.8 million to 47.5 million in the Midwest and Southeast (EPA, 1989a, p. 214). The EPA has summarized the various estimates with the conclusion that a 4°C rise in temperature could cause an increase in peak ozone concentrations by 10 per cent. The result would be to double the number of cities in exceedance of the standards from 68 to 136, causing most midsize and some small cities in the Midwest, South, and East to be added to the list of those presently in violation.

The EPA has applied its past models relating ozone concentrations to emissions of volatile organic compounds (VOC) to estimate that to offset benchmark $2xCO_2$ warming, it would be necessary to reduce VOC in the United States by 700 000 tons from a year-2000 expected base of 6 million tons. At an estimated cost of $5 000 per ton, the agency calculates that the resulting costs would amount to $3.5 billion annually (EPA, 1989a, p. 215).

Warming could affect other parts of the pollution problem as well. At higher temperatures, there is greater concentration of hydrogen peroxide (H_2O_2), the agent in the conversion of atmospheric sulphur dioxide into sulphuric acid (acid rain). Higher demand for electricity for air conditioning (discussed above) would mean greater pollution from power plants. A 10 per cent rise in electric power demand would mean a 30 per cent rise in SO_2 emissions (IPCC, 1990b, pp. 5-47). If total cloud-cover declined with global warming (see Cline, 1991a for references), there would be greater incidence of sunlight available to cause greater ozone production. The rise in water vapor associated with warming would accelerate the reaction rates of VOCs and increase production rates of ozone, hydrogen peroxide, and sulfates (EPA, 1989a, p. 207).

One way to pursue a quantitative grasp of the issue is to consider the amount being spent to abate pollution already. The EPA has estimated that, in 1990, the US economy spent $115 billion in pollution abatement, the vast bulk of it in the private sector (Roberts, 1991). Of this total, 29 per cent was for air pollution abatement. So, something like $33 billion annually was already being spent in the United States to combat air pollution. The Clean Air Act of 1990 was expected to increase annual expenses by at least $25 billion. For example, lower emissions required for automobiles were likely to cost some $500 per vehicle (about $5 billion annually), and reduction of sulphur dioxide emissions in power plants was expected to cost an additional $2 billion to $4 billion annually (Schneider, 1990; Wald, 1990). The relevant base for environmental spending on air pollution is thus in the range of some $60 billion annually.

The potential impact of global warming policy on these costs may then be visualized by asking what per centage changes in these costs might be involved. On the side of the first, direct damage avoidance, side, if a warmer climate would raise these total abatement costs by Y per cent to hold air quality standards constant, then the "benefit" of avoiding the warming amounts to $60 billion x Y per cent. On the other side, spillover benefits of greenhouse action, if lower fossil fuel burning (for example) cuts the size of the air pollution problem by Z per cent, then there is a $60 billion x Z per cent benefit from the action[31].

The IPCC "impacts" report (IPCC, 1990b) contains only two pages on air pollution, out of a volume of over 300 pages, suggesting that relatively little is known about this area. It reports very little on the impacts outside the United States. However, in view of the severe pollution problems of Eastern Europe, the former Soviet Union, and some developing countries (e.g. in Mexico City and São Paulo), there could be relatively wide geographical participation in any benefits in air pollution avoidance (and reduction) associated with policy action to abate global warming.

For much higher warming over the very long term, the effects in the area of air pollution are likely to be non-linear (rise more than proportionately with temperature). For example, although the underlying relationship of ozone concentration to temperature is approximately linear (EPA, 1989a, p. 205), the problem involves thresholds so that increased concentrations can push cites over the acceptable ceilings, and require action where none was necessary before (i.e. an infinite increase in expense from a zero base).

OVERVIEW

Table 5 presents a summary of the damage estimates developed in this study, for the case of the United States. For the very-long-term estimates, the term γ refers to the exponential pace of expansion of damage as temperature rises ($D=k[\Delta t]^\gamma$), where specific text estimates are not made[32].

The estimates here indicate that even for benchmark warming of $2xCO_2$-equivalent, damages for the United States could be relatively large. The $62 billion estimate is

Table 5. **Estimated Annual Damage From Global Warming to US Economy at 1990 Scale**
$ billion, 1990 Prices

	$2XCO_2$ (2.5°C)	Very Long Term Warming (10°C)
Agriculture	17.5	95.0
Forest Loss	3.3	7.0
Species Loss	4.0	16.0
Sea Level Rise		35.0
Dykes, Levees	1.2	
Wetlands Losses	4.1	
Drylands Losses	1.7	
Electricity Requirements	11.7	67.0
Non-electric Heating	−1.3	−4.0
Human Amenity	X_a	Y_a
Human Morbidity	5.8	33.0[i]
Migration	0.5	2.8[i]
Hurricanes	0.8	6.4[ii]
Construction	+/− X_c	+/− Y_c
Leisure Activities	1.7	4.0
Water Supply	7.0	56.0
Urban Infrastructure	0.1	0.6[i]
Air Pollution		
Tropospheric Ozone	3.5	19.8[i]
Other	X_o	Y_o
Total	61.6	338.6
	+ X_a + X_m	+ Y_a + Y_m
	+ X_o +/− X_c	+ Y_o +/− Y_c

i) $\gamma = 1.25$
ii) $\gamma = 1.50$

approximately 1.1 per cent of GDP[33]. Moreover, it excludes several of the unestimated effects, and the potentially important species loss may be underestimated. The Table's estimates for damage from very-long-term warming are more striking, and on average, are approximately $5\frac{1}{2}$ times the estimates for $2xCO_2$-equivalent.

The very-long-term damages require appropriate scaling. Most of the effects would rise proportionately with GDP, and some would rise more than proportionately (e.g. species loss, if existence value to society is an income-elastic service). The long-term damages are thus on the order of at least 6 per cent of GDP.

Conversion of the damages in Table 5 into "avoidance benefits" requires a sense of the portion of damage that could be avoided with an aggressive anti-greenhouse policy. In the IPCC scenario for "accelerated policies," radiative forcing by the year 2025 falls from 4.59 wm^{-2}, slightly more than the amount associated with doubling of carbon dioxide equivalent, to 3.52 wm^{-2}. Warming is linear in radiative forcing. The IPCC radiative forcing for business as usual amounts to warming commitment of 2.6°C by 2025; the commitment is 2.0°C even under "accelerated policies". One reason is that a substantial portion of the radiative forcing above the preindustrial base (1765) has already occurred: it stood at 2.45 wm^{-2} in 1990 (IPCC, 1990a, pp. 58-61), corresponding to a warming commitment of 1.4°C above pre-industrial levels. On this basis, more than half of the 2.5°C warming from benchmark $2xCO_2$-equivalent is already "in the pipeline".

Damage is more than linear in temperature increase. If we use $\gamma = 1.33$ on average, the reduction in the 2025 commitment to global warming from 2.6°C under IPCC "business as usual" to 2.0°C reduces the annual US damage from \$62 billion ($+\Sigma X$) to \$46 billion[34]. The net benefit from accelerated policies at the usual benchmark thus collapses to only \$16 billion annually. Moreover, it is unclear that even this estimate makes full allowance for adaptation possibilities.

For the very-long-term, as noted above, the "accelerated policies" strategy should make it possible to hold eventual global warming to only 2.5°C, rather than 10°C or more. On this basis, the net benefit from aggressive policy action would essentially equal the difference between the second and first columns in Table 5. This net gain would amount to \$277 billion at 1990 US economic scale, or approximately 5 per cent of GDP.

The thrust of this examination is twofold. First, it highlights the sobering implication that much of the damage usually considered is already unavoidable. Second, it emphatically underscores the importance of the long-term analysis. If all that were in store were the damages typically associated with benchmark $2xCO_2$ warming, there would be considerable basis for largely abandoning attempts at abatement on grounds that a large portion of the damage is unavoidable. Attention would then appropriately focus on adaptation. But if the much larger stakes in the very-long-term are valid, there are correspondingly much stronger grounds for abatement policy action.

RESEARCH NEEDS

This finding, in turn, suggests that perhaps the single most important need for research on greenhouse policy is to identify the prospective damages over the very-long-term, on the order of 250-300 years. The scientific community simply has not made these estimates. There are a few GCM runs for $4xCO_2$ (or even $8xCO_2$), but because of ample availability of fossil fuels (at least enough for $7-8xCO_2$ at economical prices; Cline, 1990*b*) *and* the fact that total radiative forcing including other trace gases is persistently about 1.45 times that from carbon dioxide alone (IPCC, 1990*a*, p. 61), warming over the very long term can reach the equivalent of as much as $15xCO_2$ (Cline, 1990*e*)[35].

The furthest out the scientific community has yet been prepared to venture is to the year 2100. For that date, the IPCC report sets radiative forcing under business as usual at 9.9 wm^{-2}, which corresponds to 5.7°C warming (at $\lambda = 2.5$). In part, economists may be responsible for the dearth of analyses at more distant horizons, because of the well-known impact of time discounting in turning very-distant estimates into negligible amounts. Economists need to emphasize to the scientists that the very-long-term estimates are necessary to appropriate policy formation, regardless of what they have heard before.

Correspondingly, a second area for research is conceptual analysis on proper discounting for the type of problem involved in global warming. The reasons for favoring the use of low discount rates were outlined earlier. However, much more complete analysis is needed, including examination of marginal utility paths under plausible assumptions of rising per capita income.

Third, there is a large research agenda merely in filling in the missing cells in the damage matrices outlined in the first section of this paper. Even for the United States, economic quantitative estimates are usually absent, and must be inferred from physical estimates. For most other countries, even the physical impact estimates are vague or non-existent. The same applies with much greater force to estimates for the very-long-term. In filling the matrix cells, a great deal of effort will be required and warranted for the developing countries in particular. Estimates for these countries are scant. Moreover, the very-long-term estimates for these countries will warrant even greater attention, as these countries will comprise a far greater share of the world's population (and, hopefully, GNP) than today (or in 2050).

Fourth, there is a need to "obtain orders of magnitude" estimates for some of the effects typically considered to be unmeasurable, such as the valuation of species loss. The most promising means for this objective is the use of survey techniques of public attitudes (i.e. "contingent evaluation").

NOTES

1. The author gratefully acknowledges partial support from the Organization for Economic Cooperation and Development. Cynthia Rosenzweig, Norman J. Rosenberg, and participants in an OECD seminar also provided helpful comments.

2. Consider, for example, Bangladesh. In Figure 1, that country is located primarily in the grid of row 18 and column 108. The boundaries spill over into neighboring grids, so grid size *is* often small enough to represent the country.

3. For example, the IPCC examines Central North America, 35-50°N, 85-105°W. This area corresponds to grid rows 12-15 and grid columns 30-37 in Figure 1, and thus to 32 grids.

4. Note that, for some countries, in at least some time periods, the initial entry in the damage matrix ($\mathbf{D}_{b,t}$) may be negative. Thus, in Iceland, the consequences of some warming could be beneficial on balance. Correspondingly, with lesser warming under policy "p", the country's entry in the benefit matrix ($\mathbf{B}_{p,t}$) would be negative as well.

5. "Commitment" is used in greenhouse discussion, and will be used here, to denote the date by which the trace gas atmospheric content has reached the level in question. The actual date of consequential global warming tends to be at least three decades later, because of thermal ocean lag.

6. This approach requires allocating the adaptation costs accross the specific economic effect categories identified for greenhouse damage. This allocation is evident for such areas as agricultural production, where the research costs for heat-resistant cultivars (for example) may be incorporated. It is less evident for adaptations in more general areas. One is the movement of populations. The most appropriate categories for allocating these costs would be the human amenity life/morbidity, and perhaps leisure activity, groupings.

7. The question asked might be along the following lines: "Suppose you have a descendant living in the year 2200. At constant prices, how many dollars would it be appropriate to take away from him (her) to make you better off by $1? How many cents would you be willing to give up to make your descendant better off by one dollar?"

8. Alternative approaches emphasising intergenerational equity will also tend to lead to a low discount rate for very-long-term analysis.

9. $CO_2 + H_2O \rightarrow CH_2O + O_2$ is photosynthesis. The reverse action is respiration.

10. US Department of Agriculture experiments in Arizona found that cotton grown in open-air fields with enhanced carbon dioxide levels showed large yield increases (Norman J. Rosenberg, by communication).

11. This calculation remains almost unchanged if account is taken of the faster phaseout of CFCs. If the CFC/HCFC radiative forcing in the IPCC Report is taken from the "accelerated policies" scenario, rather than the "business as usual" one, total radiative forcing by 2025 only drops from 4.59 wm^{-2} to 4.50 wm^{-2}. (IPCC-I, Table 2.7, p. 61).

12. The corresponding breakdown between consumers and producers *without* carbon fertilization was as follows: consumers, -$7.3 billion to -$37.5 billion; producers, +1.5 billion to +3.5 billion. *With* carbon fertilization, the breakdown becomes: consumers, -$10.3 billion to +$9.4 bil-

lion; producers, +0.6 billion to +$1.3 billion. In each case, the GFDL model shows the less-favorable results. The two models show comparable average temperature increases (GISS, 4.5°C; GFDL, 5°C), but GFDL shows serious reductions in precipitation, whereas GISS shows moderate increases in most of the five regions considered. (EPA, 1989, Appendix C, Vol. 1, pp. vi-vii).

13. Note also that the carbon-eqivalent-'doubling' represented by the approach here is above the preindustrial level (280 ppm) used by the IPCC, rather than the 1958 level (330 ppm) used by the EPA.

14. $= (^2/_3) \{ [-5.9 - 33.6]/2 \} + (^1/_3) \{ [+10.6-9.7]/2 \}$

15. The estimates do not appear to incorporate valuation changes on consumer surplus.

16. For the United States, the Table entries are based on the agricultural losses estimated above, expressed as a per cent of base output. For Australia, China, and Japan, the magnitudes are simply postulated, pending final results of the EPA international study.

17. Similarly, one does not expect a major reduction in a livestock herd to call forth higher slaughter rates; the calculus of the slaughter (extraction) rate remains, as before, a function of the interest rate and the ratio of present to future price. Over the time horizon here, future prices would not fall very fast, so the extraction rate would be unlikely to rise.

18. Thus, Nordhaus (1990) merely refers to such losses as "small".

19. The resulting extent of forest product loss ($^2/_3$ x 40 per cent = 26 per cent) would be approximately the same as identified by Binkley (1987), who estimates that would cause income from timber sales to decline by 20 per cent in the East and 26 per cent in the West by 2030, as cited in Regens, Cubbage and Hodges (1989), p. 308.

20. Smith (1991) provides the following breakdown of US forests (million hectares): "Unmanaged Natural", 120; "Managed Natural", 190; "Suburban", 80; "Urban", 40; "Intensively Managed Plantation", 20.

21. Some studies have suggested that both the carbon fertilisation effect on forests and the potential for the increased incidence of weeds, pests, and diseases may be significant in other countries. (See, for example, New Zealand Ministry of the Environment, 1990). No attempt has been made here to quantify these factors, mainly because of their uncertain scientific bases. However, it is noted in passing that they may be significant in some cases.

22. Moreover, in some northern countries, such as Finland, the value of forestry production considerably exceeds that of agriculture, so that forestry losses could dominate agricultural gains.

23. He notes that this figure is twice the amount used by Nordhaus (1990), based on the earlier EPA estimates.

24. 6 440 sq. mi. X 640 acres/sq. mi. X $1 000/acre.

25. In the Titus approach, wet and dryland losses account for 40 per cent of damage costs, and coastal defence amounts to 60 per cent. In the approach here, the composition is 75 per cent and 25 per cent, respectively.

26. This central estimate is several times as high as Nordhaus' $1.65 billion annual cost at 1981 prices, even though the Nordhaus study is based on the draft version of the same EPA study used here. Considering that the lowest figures in the EPA Table are $3 billion, there is no evident way the Nordhaus estimate could be so low, unless perhaps the annual costs were misinterpreted as capital costs. Note that, if this occurred, and the $36 billion midpoint capital costs for 2010 are added, and a 6 per cent discount rate is applied (i.e. the Nordhaus rate), and prices are converted to 1981 levels, the result would be $2.0 billion annually – close to the Nordhaus range. However, this approach would have confused capital and annual costs, and would have applied a 2010 benchmark, rather than the more appropriate 2055 level.

27. Rob Coppock, personal communication, 26 March 1991.

28. This calculation makes no allowance for the possibility that, subsequently, the taxes paid by the immigrant would exceed his share of infrastructure costs. Precisely because higher-skilled immigrants are likely to have high incomes and above-average taxes, immigration quotas tend to favor them over lower-skilled applicants. For illegal immigrants, however, skills (and earning potential) tend to be lower, and evasion of formal taxation tends to be higher. For both reasons, the taxes they pay might never come to equal their share in infrastructure costs.

29. 338 bgd X 65 days = 123 370 bgd = 123.4 g X 10^{12}. One acre-foot = 326 000 g. Annual withdrawals = $(123.4/326)$ X 10^{12-3} g = 0.378 g X 10^9.

30. Some of the effects here are for $2XCO_2$, and thus, within a time horizon of 50 years for effective warming. Others refer to a one meter sea level rise, and, therefore, a 100-year time scale.

31. There is an interaction term here that must be handled carefully. The policy action would reduce the base for the pollution problem, thereby reducing the benefit Y per cent X $60 billion. In the extreme, if the greenhouse policy action eliminated all air pollution problems, the base $60 billion would shrink to zero. The second, spillover, benefit would equal Z per cent X $60 billion = 100 per cent X $60 billion = $60 billion, but it would be double-counting at that point to attribute an additional X per cent X $60 billion for the further damage avoided.

32. Note, however, that the leisure activity estimate rises less than linearly in the long-term estimate, because the ski industry is already cut by more than half in the $2XCO_2$ scenario.

33. This estimate is approximately 7 times the direct estimate of Nordhaus: $6.6 billion at 1981 prices. However, it is consistent with his "medium damage" estimate, placed at 1 per cent of GDP, to allow for unmeasured effects in his direct estimate (Nordhaus (1990).

34. From $D = k(\Delta t)^\gamma$ and $D = 58$ at $\Delta_t = 2.5$, with $\gamma = 1.33$, we have $k = 17.2$, and $D = 43$ at $\Delta_t = 2.0$.

35. The physical reason is partly that radiative forcing rises approximately linearly with the other trace gases, which are at much less saturated concentrations, but only logarithmically in carbon dioxide.

REFERENCES

AUSUBEL, Jesse H. (1990). *Conventional Wisdom About Impacts of Global Change.*New York: Rockefeller University, Mimeograph, 4/1/90.

BATIE, Sandra S. and Shugart, Herman, H. "The Biological Consequences of Climate Changes: An Ecological and Economic Assessment", in Rosenberg *et al.* (1989*b*), pp. 121-132.

BROOKSHIRE, David S., Thayer, Mark A., Schulze, William D., and d'Arge, Ralph, C. (1982). "Valuing Public Goods: A Comparison of Survey and Hedonic Approaches", *American Economic Review,* March, pp. 165-177.

CHANGE (1990). "Newsletter on Climate Change From the Netherlands", April.

CLINE, William R. (1989). *Political Economy of the Greenhouse Effect,* Washington: Institute for International Economics, August.

————— (1990*a*). *Global Warming and the Costs of Hurricane Damage,*Washington: Institute for International Economoics, July.

————— (1990*b*). *Economic Stakes of Global Warming in the Very Long Term,*Washington: Institute for International Economics, November.

————— (1990*c*). *Grid Analysis in Global Warming Studies: A Methodological Note,* Washington: Institute for International Economics, November.

————— (1990*d*). *Greenhouse Restraints: Delaying the Inevitable At High Costs?* Washington: Institute for International Economics, December.

————— (1990*e*). *Greenhouse Gas Emissions and Global Warming: Parameters and Timetables,* Washington: Institute for International Economics, December.

————— (1991*a*). "Scientific Basis for the Greenhouse Effect", *Economic Journal,* No. 100, September, forthcoming.

————— (1991*b*). *Welfare Effects of Agricultural Yield Reductions from Global Warming,* Washington: Institue for International Economics, March.

————— (1991*c*). *A Note on Time-discounting for Consumption,* Washington: Institute for International Economics, April.

COMMERCE DEPARTMENT, USA (1990). *US Industrial Outlook 1990: Prospects for Over 350 Industries.* Washington: US Commerce Dept.

CONGRESSIONAL BUDGET OFFICE (1990). *Carbon Charges as a Response to Global Warming: The Effects of Taxing Fossil Fuels.* Washington: CBO, August.

COUNCIL OF ECONOMIC ADVISORS (CEA) (1991). *Economic Report of the President.*Washington: CEA, February.

COUNCIL ON ENVIRONMENTAL QUALITY (1991). *Environmental Quality: 21st Annual Report.* Washington: CEQ.

CROPPER, Maureen L. and Oates, Wallace (1990). *Environmental Economics: A Survey.* Washington: Resources for the Future, Discussion Paper QE90-12, January.

DARMSTADTER, Joel (1991). *The Economic Cost of CO₂ Mitigation: A Review of Estimates for Selected World Regions,* Resources for the Future Discussion Paper no. ENR91-06, Washington, January.

EDMONDS, Jae and Barns, David W. (1990). *Factors Affecting the Long-term Cost of Global Fossil Fuel CO₂ Emissions Reductions,* Washington: Pacific Northwest Laboratory.

EMMANUEL, Kerry, A. (1987). "The Dependance of Hurricane Intensity on Climate", *Nature,* 326:2, April, pp. 483-485.

ENERGY INFORMATION ADMINISTRATION (EIA) (1990). *Energy Consumption and Conservation Potential: Supporting Analysis for the National Energy Strategy,* Washington: US Department of Energy.

ENVIRONMENTAL PROTECTION AGENCY (EPA) (1989*a*). *The Potential Effects of Global Climlate Change on the United States,* Joel B. Smith and Dennis Tirpak (eds.). Washington, DC.

———— (1989*b*) *Appendix C: Agriculture,* Vols. 1 and 2.

———— (1990). *Progress Reports on International Studies of Climate Change Impacts (Draft).* Washington: EPA, November 6.

EVANS, Daniel J *et al.* (1991). *Policy Implications of Greenhouse Warming.* Washington: National Academy Press.

FLAVIN, Christopher (1989). *Slowing Global Warming: A Worldwide Strategy,* Worldwatch Paper 91, Washington: October.

GLANZ, Michael H. (1990). "Assessing the Impacts of CLimate: The Issue of Winners and Losers in a Global Climate Change Context", in Titus (ed.) *Changing Climate and the Coast,* EPA, forthcoming.

GLEICK, Peter H. (1987). "Regional Hydrologic Consequences of Increases in Atmospheric CO₂ and Other Trace Gases", *Climatic Change,* vol. 10, pp. 137-161.

GOERING, John M. (1990). *The Causes of Undocumented Migration to the United States: A Research Note.* Commission for the Study of International Migration and Cooperative Economic Development, Working Paper no. 52, July.

GOKLANY, Indur (1989). "Climate Change Effects on Fish, Wildlife, and Other DOI Programs", in Topping (ed.) (1989). *Coping With Climate Change,* pp. 273-281.

HANSEN, J. *et al.* (1989). "Regional Greenhouse Climate Effects", in Topping (ed.)(1989). *Coping With Climate Change,* pp. 68-81.

HOELLER, P., Dean, A., and Nicholaisen, J. (1990). *A Survey of Studies of the Costs of Reducing Greenhouse Gas Emissions,* OECD Working Paper no. 89, Paris.

INTERGOVERNMENTAL PANEL ON CLIMATE CHANGE (IPCC) (1990*a*). *Scientific Assessment of Climate Change: Report Prepared for the IPCC by Working Group I,* New York: WMO and UNEP, June.

———— (1990*b*). *Potential Impacts of Climate Change: Report Prepared for IPCC by Working Group II,* New York: WMO and UNEP, June.

———— (1990*c*). *IPCC First Assessment Report,* Vol. 1: Overview and Summaries, New York: WMO and UNEP, August.

IN DEPTH (1990). "Consumer Reviews for Sports Divers", vol. 5, no. 9, September.

JORGENSON, Dale W. and Wilcoxen, Peter J. (1990). *The Cost of Controlling US Carbon Dioxide Emissions,* Cambridge, Mass: Harvard University, Mimeograph, September.

KANE, Sally, Reilly, John, and Bucklin, Rhonda (1989). *Implications of the Greenhouse Effect for World Agricultural Commodity Markets,* Washington: USDA, June.

KERR, Richard A. (1991). "US Bites Greenhouse Bullet and Gags", *Science,* vol. 251, p. 868.

KIMBALL, BRUCE and ROSENBERG, Norman J. (eds.) (1990). *The Impact of CO₂ Trace Gases and Climate Change*; Madison, Wisconsin: American Society of Agronomy, Publication no. 53.

MANNE, Alan S. and Richels, Richard G. (1990a). "CO_2 Emission Limits: An Economic Cost Analysis for the USA", forthcoming, *The Energy Journal.*

———— (1990). *Buying Greenhouse Insurance,* Stanford: Stanford University, November.

MCLEAN, Dewey M. (1989). "A Mechanism for Greenhouse-induced Collapse of Mammalian Faunas", in Topping (ed.) (1989). *Coping With Climate Change,* pp. 263-267.

MEARNS, L., Katz, Richard, W., and Scheider, S., H. (1984). "Extreme High-temperature Events: Changes in Their Probabilites with Changes in Mean Temperature", *Journal of Climate and Applied Meteorology,* vol. 23, pp. 1601-1613, December.

MORGENSTERN, Richard, D. (1991). "Towards a Comprehensive Approach to Global Climate Change Mitigation", forthcoming, *American Economic Review,* May.

NATIONAL RESEARCH COUNCIL (1983). *Changing Climate,* Washington: National Academy Press.

NEW ZEALAND MINISTRY OF ENVIRONMENT (1990). "Forests", Chapter 18 in *Climate Change Impacts on New Zealand,* Wellington.

NORDHAUS, William D. (1990). *To Slow or Not To Slow: The Economics of the Greenhouse Effect,* New Haven: Yale University, Mimeograph, February.

NORDHAUS, William D. and Yohe, Garry W. (1983). "Future Paths of Energy and Carbon Dioxide Emissions", in National Research Council, (1983), pp. 87-152.

———— (1991). "A Sketch of the Economics of the Greenhouse Effect", forthcoming, *American Economic Review,* May.

OFFICE OF THE MANAGEMENT AND BUDGET (OMB) AND US DEPARTMENT OF AGRICULTURE (USDA) (1989). *Climate Impact Response Functions,* Report of a Workshop Held at Coolfont, West Virginia, September 11-14.

PARRY, Martin (1990). *Climate Change and World Agriculture,* London: Earthscan Publications.

PARRY, M. L. and Carter, T. R. (1989). "An Assessment of the Effects of Climatic Change on Agriculture, *Climatic Change,* vol. 15, pp. 95-116.

RIND, D., Goldberg, R., Hansen, J., Rosenzweig, C., and Ruedy, R. (1990). "Potential Evapotranspiration and the Likeliehood of Future Drought", *Journal of Geophysical Research,* vol. 95, no. D7, pp. 9983-10004.

RITCHIE, J. T., Baer, B. D., and Chou, T. Y. (1989). "Effect of Global Climate Change on Agriculture: Great Lakes Region", in EPA (1989b), Chapter 1.

ROBERTS, Leslie (1991). "Costs of a Clean Environment", *Science,* vol. 251, March, p. 182.

ROSENBERG, Norman J. *et al.* (1989a). *Policy Options for Adaptation to Climate Change,* Resources for the Future Discussion Paper ENR 89-05, March.

ROSENBERG, Norman J. *et al.* (1989b). *Greenhouse Warming: Abatement and Adaptation,* Washington: Resources for The Future.

ROSENBERG, Norman J. and Crosson, Pierre (1990). *Processes for Identifying Regional Influences of, and Responses to, Increasing Atmospheric CO₂ and Climate Change: The MINK Project, An Overview,* Washington: Resources for the Future, October.

ROSENZWEIG, Cynthia (1991). Personal Communication.

SCHELLING, Thomas C. (1983). "Climate Change: Implications for Welfare and Policy", in National Research Council (1983), pp. 449-486.

Appendix

IPCC ESTIMATES OF AGRICULTURAL EFFECTS OF 2XCO$_2$-EQUIVALENT

Region	Effects Anticipated
United States	Warming of 3.8°-6.3°C; soil moisture reduction of 10 per cent. Including carbon fertilization, maize yields decline 4-17 per cent in California; 16-25 per cent in the Great Plains; and 5-14 per cent in the southeast. Great Lakes experience a small potential yield increase. Wheat yields decrease by 2-3 per cent, and irrigated yields by 5-15 per cent in the Great Plains, with a northward shift of productive potential. Soybean yields fall by 3 per cent in the Great Lakes region, and by a range of 24-72 per cent in the southeast.
Canada	Warming of 3°-4°C; reduced soil moisture. Spring wheat yields decline 19 per cent nationally, despite a small increase at the northern crop frontier. Winter wheat yield declines 4 per cent. Grain, corn, barley, soybeans, hay, and potatoes all experience yield declines, except in Northern Ontario.
Mexico, Central America	Warming of 3.3°-5.4°C; rainfall changes −23 per cent to +3 per cent; will be warmer and drier; reductions in soil moisture by 10-20 per cent. Yields of maize and other rain-fed crops decline by 5-25 per cent. Heat stress and water scarcities cut irrigated yields. Possibility of enhanced soil erosion, due to more intensive rainfall.
Brazil	The Northeast will be the most vulnerable. Increased rainfall likely to be insufficient to compensate higher evapotranspiration. Could cause severe yield declines. More precipitation in centre-west could raise productivities of soybeans, maize, and in the south, wheat. Reduced frequency of frosts will be favorable for citrus and coffee production.
Argentina, Chile and North Andes	Rainfall expected to increase in presently-moist areas, and to decrease in semi-arid areas. Increase of evapotranspiration, drying of pampas, less productive cattle-raising. In Chile, with the influence of the ocean, increased winter precipitation provides greater offsets to increased evapotranspiration. Cultivation limits rise by 200 meters/°C in the high Andes.

Northern Europe	Warming 4°C; wetter. Fennoscandia benefits the most in the world from warming. Spring wheat yields increase by 10 per cent in the south of Finland, 20 per cent in the central areas, and more in the north. Iceland's grazing capacity more than doubles.
Northwest Europe	In Maritime areas, grass and potato yields rise with the higher growing season temperatures (UK potato yields increase by as much as 50-57 per cent). Grain and maize production extends several hundred km north of present limits in the UK. Wheat yields decline, despite carbon enrichment.
Southern Europe	Warming 4°C; annual rainfall reduction by more than 10 per cent. In Mediterranean regions, quite substantial decreases in productive potential. Boimass potential in Italy could fall by 5 per cent; in Greece 36 per cent. GCMs show a striking contrast between Northern Europe gains and Southern European losses. Northward shift of EEC agriculture is likely.
Alpine Europe	Climatic limits to cultivation rise 450-650 m.
Eastern Europe and former USSR	In the Leningrad region, rye yields increase until 2010, but decline thereafter, due to soil leaching and erosion. Winter wheat and maize yields increase, but barley, oats, potatoes and green vegetables decline. Perm area spring wheats could rise by 20 per cent. Saratov Region (50°N): yields rise if rainfall increases, otherwise they fall. With the low variant of rainfall, winter wheat yields are −12 per cent in Donetz-Dnieper region; −8 per cent in the southwest; −5 per cent in the central and Baltic Regions; and −7 per cent in some northern areas.
Middle East	Warming of 3.5°C; possible declines in rainfall. Experiments for Israel suggest up to a 40 per cent reduction in wheat yields.
Africa/Mahgreb	Warming of 1.5°C; evapotranspiration increases more than 10 per cent; decreases in river flows more than 10 per cent. Irrigable areas decline.
West and Northwest Africa	Increased summer rainfall, but evapotranspiration also increases. In some areas, water availability may rise. Northward extension of locusts. Possible erosion, flooding in mountain regions (e.g. Ethiopia).
East and South Africa	Depends crucially on changes in precipitation. In the current 10 per cent driest years, Kenya's yields decline by 30-70 per cent. More research is required on likely changes in rainfall frequency.
China	Stronger summer monsoons in already-rainy areas; risk of increased flooding in southern regions. With the increased rainfall, rice, maize, and wheat yields will rise by 10 per cent, but maize yields in eastern and central regions will fall by ³/₄ per cent, in the absence of higher moisture availabilites.
India	Warming reduces wheat yields by 10 per cent. If sufficient increases in rainfall occur, rice yields could rise by 7 per cent. Sorghum yields will fall because of premature development and reduced grain filling.

Japan	Rice yields rise in the north (Hokkaido) by 5 per cent, and nationally by 2-5 per cent. Maize, soybean yields rise by 4 per cent.
Oceania	New Zealand: grassland production rises 18 per cent for 3°C warming. Australia: increased summer rainfalls, decreases during winter months. Net primary productivity increases of 10-15 per cent at 30-40°S (where most precipitation is located); higher at lower latitudes. Dryland wheat yields decrease, but yields could increase with irrigation. Increased grass growth, but heat stress would affect sheep-grazing. Pacific Islands: Fiji Islands lose 6-9 per cent of their land area with a 1.5 m sea level rise. Increased tropical storms will affect copra production.

Source: IPCC, Potential Impacts of Climate Change, Working Group II, June 1990, pp. 2-14/2-21.

MAIN SALES OUTLETS OF OECD PUBLICATIONS – PRINCIPAUX POINTS DE VENTE DES PUBLICATIONS DE L'OCDE

Argentina – Argentine
Carlos Hirsch S.R.L.
Galería Güemes, Florida 165, 4° Piso
1333 Buenos Aires   Tel. (1) 331.1787 y 331.2391
Telefax: (1) 331.1787

Australia – Australie
D.A. Book (Aust.) Pty. Ltd.
648 Whitehorse Road, P.O.B 163
Mitcham, Victoria 3132   Tel. (03) 873.4411
Telefax: (03) 873.5679

Austria – Autriche
OECD Publications and Information Centre
Schedestrasse 7
D-W 5300 Bonn 1 (Germany)   Tel. (49.228) 21.60.45
Telefax: (49.228) 26.11.04

Gerold & Co.
Graben 31
Wien I   Tel. (0222) 533.50.14

Belgium – Belgique
Jean De Lannoy
Avenue du Roi 202
B-1060 Bruxelles   Tel. (02) 538.51.69/538.08.41
Telefax: (02) 538.08.41

Canada
Renouf Publishing Company Ltd.
1294 Algoma Road
Ottawa, ON K1B 3W8   Tel. (613) 741.4333
Telefax: (613) 741.5439
Stores:
61 Sparks Street
Ottawa, ON K1P 5R1   Tel. (613) 238.8985
211 Yonge Street
Toronto, ON M5B 1M4   Tel. (416) 363.3171
Federal Publications
165 University Avenue
Toronto, ON M5H 3B8   Tel. (416) 581.1552
Telefax: (416)581.1743
Les Éditions La Liberté Inc.
3020 Chemin Sainte-Foy
Sainte-Foy, PQ G1X 3V6   Tel. (418) 658.3763
Telefax: (418) 658.3763

China – Chine
China National Publications Import
Export Corporation (CNPIEC)
P.O. Box 88
Beijing   Tel. 44.0731
Telefax: 401.5661

Denmark – Danemark
Munksgaard Export and Subscription Service
35, Nørre Søgade, P.O. Box 2148
DK-1016 København K   Tel. (33) 12.85.70
Telefax: (33) 12.93.87

Finland – Finlande
Akateeminen Kirjakauppa
Keskuskatu 1, P.O. Box 128
00100 Helsinki   Tel. (358 0) 12141
Telefax: (358 0) 121.4441

France
OECD/OCDE
Mail Orders/Commandes par correspondance:
2, rue André-Pascal
75775 Paris Cédex 16   Tel. (33-1) 45.24.82.00
Telefax: (33-1) 45.24.85.00
or (33-1) 45.24.81.76
Telex: 620 160 OCDE
Bookshop/Librairie:
33, rue Octave-Feuillet
75016 Paris   Tel. (33-1) 45.24.81.67
(33-1) 45.24.81.81
Librairie de l'Université
12a, rue Nazareth
13100 Aix-en-Provence   Tel. 42.26.18.08
Telefax: 42.26.63.26

Germany – Allemagne
OECD Publications and Information Centre
Schedestrasse 7
D-W 5300 Bonn 1   Tel. (0228) 21.60.45
Telefax: (0228) 26.11.04

Greece – Grèce
Librairie Kauffmann
Mavrokordatou 9
106 78 Athens   Tel. 322.21.60
Telefax: 363.39.67

Hong Kong
Swindon Book Co. Ltd.
13 - 15 Lock Road
Kowloon, Hong Kong   Tel. 366.80.31
Telefax: 739.49.75

Iceland – Islande
Mál Mog Menning
Laugavegi 18, Pósthólf 392
121 Reykjavik   Tel. 162.35.23

India – Inde
Oxford Book and Stationery Co.
Scindia House
New Delhi 110001   Tel.(11) 331.5896/5308
Telefax: (11) 332.5993
17 Park Street
Calcutta 700016   Tel. 240832

Indonesia – Indonésie
Pdii-Lipi
P.O. Box 269/JKSMG/88
Jakarta 12790   Tel. 583467
Telex: 62 875

Ireland – Irlande
TDC Publishers – Library Suppliers
12 North Frederick Street
Dublin 1   Tel. 74.48.35/74.96.77
Telefax: 74.84.16

Israel
Electronic Publications only
Publications électroniques seulement
Sophist Systems Ltd.
71 Allenby Street
Tel-Aviv 65134   Tel. 3-29.00.21
Telefax: 3-29.92.39

Italy – Italie
Libreria Commissionaria Sansoni
Via Duca di Calabria 1/1
50125 Firenze   Tel. (055) 64.54.15
Telefax: (055) 64.12.57
Via Bartolini 29
20155 Milano   Tel. (02) 36.50.83
Editrice e Libreria Herder
Piazza Montecitorio 120
00186 Roma   Tel. 679.46.28
Telex: NATEL I 621427
Libreria Hoepli
Via Hoepli 5
20121 Milano   Tel. (02) 86.54.46
Telefax: (02) 805.28.86
Libreria Scientifica
Dott. Lucio de Biasio 'Aeiou'
Via Meravigli 16
20123 Milano   Tel. (02) 805.68.98
Telefax: (02) 80.01.75

Japan – Japon
OECD Publications and Information Centre
Landic Akasaka Building
2-3-4 Akasaka, Minato-ku
Tokyo 107   Tel. (81.3) 3586.2016
Telefax: (81.3) 3584.7929

Korea – Corée
Kyobo Book Centre Co. Ltd.
P.O. Box 1658, Kwang Hwa Moon
Seoul   Tel. 730.78.91
Telefax: 735.00.30

Malaysia – Malaisie
Co-operative Bookshop Ltd.
University of Malaya
P.O. Box 1127, Jalan Pantai Baru
59700 Kuala Lumpur
Malaysia   Tel. 756.5000/756.5425
Telefax: 757.3661

Netherlands – Pays-Bas
SDU Uitgeverij
Christoffel Plantijnstraat 2
Postbus 20014
2500 EA's-Gravenhage   Tel. (070 3) 78.99.11
Voor bestellingen: Tel. (070 3) 78.98.80
Telefax: (070 3) 47.63.51

New Zealand – Nouvelle-Zélande
GP Publications Ltd.
Customer Services
33 The Esplanade - P.O. Box 38-900
Petone, Wellington   Tel. (04) 5685.555
Telefax: (04) 5685.333

Norway – Norvège
Narvesen Info Center - NIC
Bertrand Narvesens vei 2
P.O. Box 6125 Etterstad
0602 Oslo 6   Tel. (02) 57.33.00
Telefax: (02) 68.19.01

Pakistan
Mirza Book Agency
65 Shahrah Quaid-E-Azam
Lahore 3   Tel. 66.839
Telex: 44886 UBL PK. Attn: MIRZA BK

Portugal
Livraria Portugal
Rua do Carmo 70-74
Apart. 2681
1117 Lisboa Codex   Tel.: (01) 347.49.82/3/4/5
Telefax: (01) 347.02.64

Singapore – Singapour
Information Publications Pte. Ltd.
Pei-Fu Industrial Building
24 New Industrial Road No. 02-06
Singapore 1953   Tel. 283.1786/283.1798
Telefax: 284.8875

Spain – Espagne
Mundi-Prensa Libros S.A.
Castelló 37, Apartado 1223
Madrid 28001   Tel. (91) 431.33.99
Telefax: (91) 575.39.98
Libreria Internacional AEDOS
Consejo de Ciento 391
08009 - Barcelona   Tel. (93) 488.34.92
Telefax: (93) 487.76.59
Llibreria de la Generalitat
Palau Moja
Rambla dels Estudis, 118
08002 - Barcelona   Tel. (93) 318.80.12 (Subscripcions)
(93) 302.67.23 (Publicacions)
Telefax: (93) 412.18.54

Sri Lanka
Centre for Policy Research
c/o Colombo Agencies Ltd.
No. 300-304, Galle Road
Colombo 3   Tel. (1) 574240, 573551-2
Telefax: (1) 575394, 510711

Sweden – Suède
Fritzes Fackboksföretaget
Box 16356
Regeringsgatan 12
103 27 Stockholm   Tel. (08) 23.89.00
Telefax: (08) 20.50.21
Subscription Agency/Abonnements:
Wennergren-Williams AB
Nordenflychtsvägen 74
Box 30004
104 25 Stockholm   Tel. (08) 13.67.00
Telefax: (08) 618.62.32

Switzerland – Suisse
OECD Publications and Information Centre
Schedestrasse 7
D-W 5300 Bonn 1 (Germany)   Tel. (49.228) 21.60.45
Telefax: (49.228) 26.11.04
Suisse romande
Maditec S.A.
Chemin des Palettes 4
1020 Renens/Lausanne   Tel. (021) 635.08.65
Telefax: (021) 635.07.80
Librairie Payot
6 rue Grenus
1211 Genève 11   Tel. (022) 731.89.50
Telex: 28356
Subscription Agency – Service des Abonnements
Naville S.A.
7, rue Lévrier
1201 Genève   Tél.: (022) 732.24.00
Telefax: (022) 738.87.13

Taiwan – Formose
Good Faith Worldwide Int'l. Co. Ltd.
9th Floor, No. 118, Sec. 2
Chung Hsiao E. Road
Taipei   Tel. (02) 391.7396/391.7397
Telefax: (02) 394.9176

Thailand – Thaïlande
Suksit Siam Co. Ltd.
113, 115 Fuang Nakhon Rd.
Opp. Wat Rajbopith
Bangkok 10200   Tel. (662) 251.1630
Telefax: (662) 236.7783

Turkey – Turquie
Kültur Yayinlari Is-Türk Ltd. Sti.
Atatürk Bulvari No. 191/Kat. 21
Kavaklidere/Ankara   Tel. 25.07.60
Dolmabahce Cad. No. 29
Besiktas/Istanbul   Tel. 160.71.88
Telex: 43482B

United Kingdom – Royaume-Uni
HMSO
Gen. enquiries   Tel. (071) 873 0011
Postal orders only:
P.O. Box 276, London SW8 5DT
Personal Callers HMSO Bookshop
49 High Holborn, London WC1V 6HB
Telefax: 071 873 2000
Branches at: Belfast, Birmingham, Bristol, Edinburgh,
Manchester

United States – États-Unis
OECD Publications and Information Centre
2001 L Street N.W., Suite 700
Washington, D.C. 20036-4910   Tel. (202) 785.6323
Telefax: (202) 785.0350

Venezuela
Libreria del Este
Avda F. Miranda 52, Aptdo. 60337
Edificio Galipán
Caracas 106   Tel. 951.1705/951.2307/951.1297
Telegram: Libreste Caracas

Yugoslavia – Yougoslavie
Jugoslovenska Knjiga
Knez Mihajlova 2, P.O. Box 36
Beograd   Tel. (011) 621.992
Telefax: (011) 625.970

Orders and inquiries from countries where Distributors have
not yet been appointed should be sent to: OECD Publications Service, 2 rue André-Pascal, 75775 Paris Cédex 16,
France.

Les commandes provenant de pays où l'OCDE n'a pas
encore désigné de distributeur devraient être adressées à :
OCDE, Service des Publications, 2, rue André-Pascal, 75775
Paris Cédex 16, France.

OECD PUBLICATIONS, 2 rue André-Pascal, 75775 PARIS CEDEX 16
PRINTED IN FRANCE
(97 92 04 1) ISBN 92-64-13639-8 - No. 45963 1992